About Island Press

Island Press is the only nonprofit organization in the United States whose principal purpose is the publication of books on environmental issues and natural resource management. We provide solutions-oriented information to professionals, public officials, business and community leaders, and concerned citizens who are shaping responses to environmental problems.

In 1994, Island Press celebrated its tenth anniversary as the leading provider of timely and practical books that take a multidisciplinary approach to critical environmental concerns. Our growing list of titles reflects our commitment to bringing the best of an expanding body of literature to the environmental community throughout North America and the world.

Support for Island Press is provided by Apple Computer, Inc., The Bullitt Foundation, The Geraldine R. Dodge Foundation, The Energy Foundation, The Ford Foundation, The W. Alton Jones Foundation, The Lyndhurst Foundation, The John D. and Catherine T. MacArthur Foundation, The Andrew W. Mellon Foundation, The Joyce Mertz-Gilmore Foundation, The National Fish and Wildlife Foundation, The Pew Charitable Trusts, The Pew Global Stewardship Initiative, The Rockefeller Philanthropic Collaborative, Inc., and individual donors.

Mitigation Banking

Mitigation Banking

Mitigation Banking

THEORY AND PRACTICE

Edited by Lindell L. Marsh,
Douglas R. Porter, and David A. Salvesen

In cooperation with THE URBAN LAND INSTITUTE

FOREWORD BY JOHN DEGROVE

ISLAND PRESS
Washington, D.C. □ Covelo, California

ULI provided a review of the manuscript by the following ULI members: Timothy Beatley, Associate Professor, Department of Urban and Regional Planning, University of Virginia; Ted R. Brown, Vice President/General Counsel, Arvida Company; Byron R. Koste, President, WCI Communities, Inc.; Dwight H. Merriam, Partner, Robinson & Cole; Laurence M. Netherton, Vice President, Chevron Land and Development; and Paul O. Reimer, President, Reimer Associates.

Library of Congress Cataloging-in-Publication Data

Mitigation Banking: theory and practice/edited by Lindell Marsh, Doug Porter, and David Salvesen; foreword by John DeGrove.
p. cm.
Includes bibliographical references and index.
ISBN 1-55963-371-9 (pbk.)
1. Wetland conservation. 2. Wetland mitigation banking.
I. Marsh, Lindell L. II. Porter, Douglas R. III. Salvesen, David.
QH75.M525 1996
333.91'816—dc20 96-840
CIP

Printed on recycled, acid-free paper

Manufactured in the United State of America

10 9 8 7 6 5 4 3 2 1

Mitigation Banking

THEORY AND PRACTICE

Edited by Lindell L. Marsh,
Douglas R. Porter, and David A. Salvesen

In cooperation with THE URBAN LAND INSTITUTE

FOREWORD BY JOHN DEGROVE

ISLAND PRESS
Washington, D.C. ❐ Covelo, California

ULI provided a review of the manuscript by the following ULI members: Timothy Beatley, Associate Professor, Department of Urban and Regional Planning, University of Virginia; Ted R. Brown, Vice President/General Counsel, Arvida Company; Byron R. Koste, President, WCI Communities, Inc.; Dwight H. Merriam, Partner, Robinson & Cole; Laurence M. Netherton, Vice President, Chevron Land and Development; and Paul O. Reimer, President, Reimer Associates.

Library of Congress Cataloging-in-Publication Data

Mitigation Banking: theory and practice/edited by Lindell Marsh, Doug Porter, and David Salvesen; foreword by John DeGrove.
p. cm.
Includes bibliographical references and index.
ISBN 1-55963-371-9 (pbk.)
1. Wetland conservation. 2. Wetland mitigation banking. I. Marsh, Lindell L. II. Porter, Douglas R. III. Salvesen, David.
QH75.M525 1996
333.91'816—dc20 96-840
CIP

Printed on recycled, acid-free paper

Manufactured in the United State of America

10 9 8 7 6 5 4 3 2 1

Contents

Glossary

Like any emerging field, wetlands mitigation has developed its own lexicon. The vocabulary of wetlands mitigation and of mitigation banking includes awkward terms such as "out-of-kind replacement" and "single-user banks." The words "sequencing" and "creation" take on a new meaning in the context of wetlands regulation or mitigation. Some of the more common terms are defined in this brief glossary. Many are discussed in more detail in the text.

Creation Establishing wetlands on a site where they have not existed before.

Degraded wetland A wetland that no longer provides the full range of its natural functions and values, due to man-made alterations or disturbances, such as invasion by exotic species, siltation, pollution, deposit of fill or refuse, or changes in hydrology.

Enhancement Alteration of an existing wetland to increase specific functions (e.g., flooding a freshwater marsh to create a pond that will provide greater habitat for waterfowl).

Entrepreneurial banks A mitigation bank (see definition), established by a landowner and/or investor(s), in which the credits are sold on the open market to compensate for wetland losses.

Exchange ratio The ratio of the number of acres of created, restored, or enhanced wetlands that can be exchanged for a certain number of acres of wetlands lost to development. For example, if ten acres of created wetlands are required to compensate for the loss of five acres of a natural wetland, the exchange ratio is 2:1.

In-kind replacement Replacing one type of wetland with another of the same type and condition, for example, compensating for the loss of a salt marsh by creating or restoring a physically and biologically similar salt marsh, as opposed to what is awkwardly referred to as "out-of-kind" replacement (e.g., replacing a salt marsh with a bottomland hardwood swamp).

Mitigation As defined by the Council of Environmental Quality, Mitigation includes (1) avoiding the impact altogether by not taking a certain action; (2) minimizing impacts by limiting the degree of the action; (3) rectifying the impact by repairing, rehabilitating, or restoring the affected environment; (4) reducing or eliminating the impact over time by preservation and maintenance operations; and (5) compensating for the impact by replacing or providing substitute resources (40CFR Section 1508.20).

Mitigation bank The creation, restoration, or enhancement of wetlands that will be sold or exchanged to compensate for future wetland losses. Typically, the created, restored, or enhanced wetlands are designated as a bank. The value of the wetlands created, restored, or enhanced are somehow quantified and assigned credits, which can be sold or "withdrawn" to compensate for the loss of wetlands or wildlife habitat elsewhere.

Mitigation credits A unit of measure of the increase in wetland functional value achieved at a mitigation site, and therefore a unit of exchange for compensatory mitigation.

No net loss The point at which wetland losses equal wetland gains (i.e., ten acres of wetlands created to compensate for the loss of ten acres).

Off-site Not on the same parcel of land as a wetland that has been adversely affected by a particular development.

Out-of-kind replacement Compensating for the loss of one type of wetland by creating or restoring a wetland of a different kind or type (e.g., replacing the loss of a bottomland hardwood swamp with an estuarine marsh).

Restoration Reestablishing a damaged or degraded wetland to a close approximation of its predegraded condition (e.g., by eradicating exotic plants or removing fill).

Sequencing Following the mitigation steps of the Council of Environmental Quality's definition of mitigation, in numerical order, (see earlier definition), that is, avoid what you can, minimize where possible, and, after that, compensate for the damage done.

Single-user bank A mitigation bank (see earlier definition) established by one company or agency, such as an oil company or a port authority, from which it can later withdraw credits to compensate for future wetland losses resulting from its own projects.

Foreword

Slowly but surely we are finding workable and equitable ways to preserve our threatened environment in the face of continued development. The environmental laws enacted at federal and state levels from the 1960s onward were a giant leap forward in raising our sensitivity to preservation needs. Reacting to a long period of environmental degradation, the laws set forth some ambitious objectives and procedures bent on maximizing preservation of natural values, qualities, and features—certainly a laudable intention. This approach succeeded in preventing the worst abuse of wetlands and wildlife habitats in many areas across the nation, but has not reversed the net loss of resources.

Critical weaknesses in the laws and rules adopted to implement them have become apparent. Permitting procedures are reactive, operate on a case-by-case basis, and are designed to prevent the bad rather than to achieve the good. Confusing and often contradictory laws and requirements, combined with the large number of federal, state, regional, and local agencies charged with permitting, produce unpredictable results. The case-by-case approach falls short of accounting for cumulative impacts. Permitting often takes place without reference to a framework of clearly set planning goals and policies. By the mid-1980s, these weaknesses had produced an extremely adversarial process, with developers and regulators at war and neither achieving their goals in an effective or efficient manner.

The National Wetlands Forum convened by the Conservation Foundation in 1987 and chaired by Governor Tom Kean of New Jersey brought together all the relevant stakeholders involved in wetlands issues. The forum's final report focused on ending definitional confusion, consolidating permitting in fewer agencies with a strong emphasis on state responsibilities, and addressing wetlands protection on a broad, watershed basis instead of attempting to protect the resource by a reactive, single-permit approach.

In this context of a new approach to reconciling preservation needs with necessities for responding to economic and population growth, mitigation banking emerges as an important new tool. Mitigation banking allows public agencies and private landowners to replace damaged or destroyed wetlands and habitats by funding restoration or creation of like features elsewhere. Entrepreneurial banks are the newest wrinkle, offering a convenient, market-based medium by which to satisfy mitigation needs while maintaining or even enlarging the inventory of threatened resources.

In my opinion, mitigation banks offer a promising institutional mechanism that should be pieced into the environmental protection puzzle in a responsible and accountable way. The writers in this book raise many legitimate issues about mitigation banks, private or public, and whether they can be employed in a responsible way. I believe that they can, but the rules for guiding their use must evolve with experience. It is clear to me that the effort to protect isolated small pockets of wetlands has not been successful; certainly, habitat conservation is not feasible in bits and pieces. We must move to a broad, watershed-planning approach if we are ever to achieve the often-embraced goal of "no net loss" of wetlands and a similar level of protection for wildlife diversity. Mitigation banks can help in achieving that shift in focus.

A compelling illustration of the effectiveness of mitigation banks is emerging in Florida. A unique partnership between Disney World, the Florida chapter of The Nature Conservancy, and the state's environmental regulatory agency has produced a wetland mitigation bank originally funded by Disney and developed and managed by The Nature Conservancy. A watershed approach to wetland protection, the bank started with an 8,500-acre part of the watershed and has since expanded to more than 10,000 acres as other organizations have turned to The Nature Conservancy for wetland mitigation credits. The Orlando International Airport Authority, for example, has been a major purchaser of credits. Many lessons about the complexities of achieving long-term viability of wetland mitigation banks are being played out in this project.

Mitigation bankers have fought an uphill battle to gain respect in the regulatory environment. In past years they have lacked a supportive public policy framework and have operated under a cloud of apprehension and misunderstanding. The book makes the case, however, that mitigation banking can smooth the path toward reconciling environmental and development aims, moving us closer to a sustainable environment.

JOHN DEGROVE

Director, Florida Atlantic University/Florida International University Joint Center for Environmental and Urban Problems, Fort Lauderdale, Florida

1
Introduction and Overview

Lindell L. Marsh, Douglas R. Porter, and David A. Salvesen

The idea of mitigation banking was born in the 1970s, based on the need for a simpler way to mitigate the loss of wetlands caused by development projects, as required by laws such as the federal Clean Water Act (CWA) of 1972. Typically, developers had neither the expertise nor the incentive to mitigate the impacts of their projects on wetlands. Using a market approach, however, a third party ("mitigation banker") could create, restore, or enhance wetlands to create a bank of wetland credits that could be sold or conveyed to a developer, who would utilize the credits to compensate for the adverse impacts to wetlands caused by the developer's project. The banked lands would continue to be held and operated by the banker or its successor to conserve the wetlands in perpetuity (much like the perpetual-care concept associated with cemeteries), with appropriate assurances to this effect provided to the agencies. This is the standard model, often referred to as an "entrepreneurial bank."

While the idea had broad appeal, two decades later only a few entrepreneurial banks are in operation, and virtually none that is beyond the "start-up" phase. This is due in large part to several factors: our lack of understanding of the wetland ecosystems involved, the difficulty of reconciling the requirements of a variety of regulations and agencies, and an inadequate legal policy and planning framework. In addition, mitigation banking suffered from an underlying hostility between the development community and both the conservation community and a cadre of regulatory agency staffers. This hostility can be characterized as a deep institutional anger toward anything that would make development easier or that might provide a loophole in the strong environmental regulations that had been promulgated during the late 1960s and early 1970s.

Now, however, the mitigation banking concept is undergoing a renewal. While the basic idea is much the same, the concept has been modified to address prior impediments. To understand mitigation banking in today's environment, it is important to understand these impediments in a historic context.

The Troubled History of Mitigation Banking

Our system of governance is based on individual freedoms, one of which is the right to own property—land—and to choose its uses subject only to limited constraints. The system has great vitality, similar in some ways to the vitality provided by increased biodiversity, enabling a broad multitude of individual initiatives and actions—evidenced by the creativity for which our nation is known. There is, however, a dark side to the system. Individuals, corporations, and agencies undertake projects that by themselves cause relatively minor adverse impacts to the environment but together may result in substantial cumulative impacts, such as the destruction of crucial wetlands or wildlife habitat. These cumulative impacts, which economists call externalities, are not adequately accounted for or controlled under our current project-by-project system of regulation.

During the late 1960s to early 1970s, following the major economic expansion triggered by the conclusion of World War II, a series of environmental laws and regulations were adopted at the local, state, and federal levels, based on various general powers (police, commerce, treaty, etc.). These laws and initiatives profoundly changed the direction of state and national policy. They reflected a growing disenchantment with externalities of development and an antipathy that was expressed through the actions of agencies empowered to implement the regulations (an antipathy that was exacerbated by the resistance of economic development interests to change). This was the milieu in the 1970s that caused the concept of mitigation banking to be greeted with caution and even hostility.

Different levels of government attempted to protect environmental resources with separate standards and preferences, imposed through "command and control" techniques. In general, with regard to land use, environmental policy has been implemented by requiring new development to bear the burden of past and current adverse impacts. However, due to the historic project-by-project approach, little groundwork was laid for establishing the linkage between individual actions and their cumulative effects, and for understanding the ecosystems involved. Equally important, the legal and institutional linkages and frameworks to relate individual project impacts with relevant ecosystems as a whole had yet to be developed. Conservationists and regulatory agencies were troubled by the limited scientific knowledge available. Proposals for mitigation banking raised profound questions regarding the value and function of wetlands, wildlife, and habitat that were difficult to answer with any acceptable degree of certainty. How could wetlands be created, restored, and maintained? Would created or restored wetlands be as valuable as those lost, particularly if they were in a different and perhaps more distant location? How could the impacts and proposed compensation be measured and compared accurately? As a result of these uncertainties, the concept of compensating for wetland losses with acquisition and creation or restoration of wetlands was constrained by the promulgation of the Environmental Protection Agency's (EPA) Section 404(b)(1) guidelines in 1975 and their revisions in 1980 (see box on pp. 4–5). Within the Section 404 regulatory program, the guidelines skewed mitigation policy away from the compensation concept toward a preference for preservation in place. Policy statements within the guidelines reflected the skepticism among resource agencies and the environmental community about whether man-made wetlands could adequately replace the functions performed by natural ecosystems. The result was a wide spectrum of mitigation ideas promulgated by development interests, as well as an institutional reluctance to make the scientific judgments on wetland functions and values, and to invent the institutional elements that would support mitigation banking.

Meanwhile, development interests were clambering, in some cases desperately, to find a way through the regulatory maze that the regulations and guidelines had created. Developers had little understanding of, and often little patience for, environmental concerns. Still, some successful wetland restoration projects in the mid- to late 1970s began to reduce the general skepticism about the effectiveness of compensatory mitigation in reducing adverse environmental impacts. In the early 1980s, increasing numbers of Section 404 permit applications included proposals for compensatory mitigation. Throughout the late 1970s and 1980s, the U.S. Army Corps of Engineers (Corps) generally considered off-site mitigation as a legitimate mitigation solution, in part because its experience demonstrated that wetlands preserved on development sites could be damaged or destroyed by other impacts, such as siltation and

A Primer on Wetlands Mitigation Banking*

Under Section 404 of the Clean Water Act, a permit from the U.S. Army Corps of Engineers (Corps) is usually required to develop in wetlands. The Corps administers the 404 permit program, following guidelines established by the Environmental Protection Agency (EPA). In general, the Corps will grant a permit on the condition that an applicant, in the following sequence, (1) takes all practicable steps to avoid adverse impacts to wetlands, (2) minimizes unavoidable damage to wetlands, and (3) compensates for permanent destruction of wetlands by creating a new wetland or by restoring a degraded one. Together, these three measures are referred to as mitigation.

Where compensation is required, both the Corps and EPA prefer that the same kind of wetland be created on the same site as the one being filled, preferably as close to the destroyed wetland as possible and connected to the same water source. This, the agencies surmise, will increase the likelihood that the natural functions and values (wildlife habitat, stormwater retention, etc.) lost when a wetland is destroyed will be replaced adequately by a new wetland. For example, suppose a developer applies for a Section 404 permit to construct an office/retail complex on a 15-acre site, half of which is occupied by a freshwater marsh. After following steps (1) and (2), the developer, nonetheless, still needs to fill four acres of wetlands for the project to be feasible. To compensate for the loss of those 4 acres, the developer is required to construct a 4-acre freshwater marsh somewhere on the 15-acre site, although it is not uncommon for the Corps to require that a larger wetland, say 6 or 8 acres, be created. If the creation of wetlands on-site is infeasible, the Corps will allow the wetlands to be constructed off-site.

Instead of requiring developers to create and maintain wetlands, a task they are not well trained for, it may be preferable, both economically and ecologically, to allow them to purchase a wetland created and maintained properly by someone else (e.g., through a mitigation bank). Although there are many variations, mitigation banking generally works something like this: a degraded wetland is purchased and restored by one party, such as a government agency or an investor; this site becomes the bank. The values of the restored wetland are somehow quantified and used as "credits" that can later be withdrawn, at a price, to compensate for unavoidable wetland fills elsewhere. Viewed broadly, however, mitigation banking encompasses a variety of mechanisms to offset or compensate for future losses of wetlands or wildlife habitat, including:

1. Single-Owner/User Banks—the most common form. Usually established by a large company or public agency whose future development plans call for filling numerous small wetlands over several years. The company or agency may create a large wetland from which it can later use or withdraw "credits" as compensation for (future) wetland fills, rather than create a small wetland at each site on a case-by-case basis .

---

*Adapted from David A. Salvesen, "Banking On Wetland," *Urban Land,* June 1993.

2. Entrepreneurial Banks—very few exist. Similar in principle to single-owner banks except that the bank is established by a landowner and/or investor and the credits can be purchased by anyone.
3. Joint Projects—common in California. Usually do not involve establishing a bank of credits for future use. Typically, a consortium of developers agree to fund a mitigation project to compensate for specific, future losses of wetlands or endangered species habitat.

---

other hydrologic alterations. Many Corps field staff preferred to allow wetland fills in exchange for an agreement that off-site mitigation sites would be managed or restored to maintain wetland functions and values.

The Environmental Protection Agency, on the other hand, interpreted the Section 404 (b)(1) guidelines to require that applicants follow a sequence of priorities in determining appropriate mitigation measures. Based on the high failure rate of attempted wetlands creation and restoration projects, EPA required evidence that applicants had avoided and minimized wetland impacts to the maximum extent possible before entertaining off-site mitigation proposals.

The conflicting approaches of these two agencies created tensions and delays. Mitigation banking proposals encountered predictable and often insurmountable difficulties posed by public agency responses and by business concerns regarding investment risk. In practical terms, mitigation banking ideas were met with lengthy delays in processing permit approvals, often capped with conflicting agency requirements. Frequently, these requirements were based on conservative approaches, such as narrowly defined "in-kind" and "geographically proximate" replacement criteria, or requirements that the bank be fully developed and proven for a significant period of time before the agencies would recognize any credits of the bank.

To potential entrepreneurial mitigation bankers, the risk attendant with these requirements was unacceptable. In addition, there were no guarantees that the agencies involved in the process would cooperate to make the banks successful. Staffs could change, and with them agreements and understandings. New staff members might interpret the "in-kind" requirement differently or more narrowly, excluding potential credit purchasers and encouraging them to use, in effect, a different bank or off-site mitigation source. Faced with these uncertainties, potential bankers concluded that the up-front costs and risk outweighed the potential reward and declined to participate.

The one exception was the "single user bank," where a highway agency or oil producer, for example, would agree to develop a mitigation bank in advance of a series of development projects or project phases that would occur in wetlands. In this case, the potential benefits outweighed the risks, in large part because of the ongoing relationships between the parties (such as interagency relationships), the ability to rely upon relatively general mitigation standards, and the geographically extensive impacts anticipated.

In 1990, when the Corps and EPA issued a Memorandum of Agreement (MOA), and then again in 1993 (see discussion of MOAs in chapter 7) when the two agencies issued joint guidance on mitigation banking, the regulatory situation became somewhat clearer. Although the MOA establishes preferences for avoidance and minimization of impacts and for on-site mitigation, off-site mitigation banks are considered acceptable instruments in some circumstances. The joint guidance increased the flexibility with which the Corps and EPA apply the Section 404 (b)(1) guidelines and specifically states that "the agencies' preference for on-site, in-kind mitigation does not preclude the use of mitigation banks. . . ."

In November 1995, the Corps and EPA, along with the Fish and Wildlife Service, National Marine Fisheries Service, and the Natural Resources Conservation Service (NCRS, formerly the Soil Conservation Service) issued policy guidance that further clarified the role of mitigation banks in compensating for adverse impacts to wetlands and other aquatic resources. The 1995 joint federal guidance reiterated the agencies' preference for on-site mitigation, but stated that mitigation banks may be used where on-site mitigation is not practicable or where the "use of a mitigation bank is environmentally preferable." (See chapter 3 for a discussion of the joint federal guidance, see federal mitigation banking guidance in the appendix.)

A Shift in Paradigm

It appears that a major shift is occurring in the governance framework relating to mitigation banking from that established in the late 1960s. First, fragmented efforts that led to inefficiencies and conflicts are giving way to commitments to collaborative actions among public agencies as well as between the public and private sectors. The trend toward collaboration is reflected in the private sector by a movement toward more horizontal management structures, characterized by management by principles and values, or "partnering." These structures emphasize open and honest communication, the promotion of trust and commitment to work in good faith toward the reconciliation of conflicts (as opposed to compromise, which implies a lowering of standards), with the objective of achieving a solution that addresses all concerns to the extent practicable. The trend is fueled by an increasing shift toward a broad national consensus that wetlands and wildlife conservation are appropriate societal values and that project-by-project mitigation has failed to achieve its objectives.

Second, in addition to the emergence of groups of regulators, developers, and consultants who have grown more accustomed to working together—developing a vocabulary, protocols, and understandings, and a degree of trust in mutual commitments, assumptions, and objectives—planning processes have been developed to enhance collaborative efforts. The conceptual model for these processes is the National Environmental Policy Act (NEPA), which establishes procedures for exploring a proposed action, together with possible alternatives and their impacts, with the constituency of concerned organizations, interests, and agencies.

Prior to 1965, environmental and land use planning reports prepared by agency staff or consultants often were disregarded, with some exceptions (e.g., plans for some river basins and national parks and forests). In part, such plans did not reflect the many factors that would ultimately affect the public policy decisions that would follow. During the late 1960s and early 1970s, plans began to be the repository of public policy decisions. In 1973, Florida adopted much of the American Law Institute *Model Land Development Code* and California had provided for "specific plans" in addition to general (comprehensive) plans. Plans were also called for by New York in connection with its proposed Adirondack Park (including private and public lands) and by California for its coastal zone, Lake Tahoe, and San Francisco Bay. New Jersey later developed a plan for the conservation of the 1.2-million acres within the Pinelands. And the federal government enacted the Coastal Zone Management Act, which fostered coastal plans by the various coastal states.

During the late 1970s and 1980s, the use of plans as repositories of public policy and the process concepts contained within the NEPA converged. Major large-scale planning efforts occurred for special areas, such as Grays Harbor Washington (which ultimately failed because of defects in process and the legal framework) and later, the Chesapeake Bay. The prime example, however, is the amendment to the federal Endangered Species Act (ESA) in 1982, which institutionalized the use of habitat conservation plans (HCPs) for this purpose. The amendment contemplates that the "constituency of interests" will collaborate to develop a habitat conservation plan and reconcile the competing concerns. Natural Community Conservation Plans (NCCP), which are large-scale HCPs, are now being prepared for virtually all of the undeveloped urbanizing areas of Southern California. Similar efforts are underway in the Northwest, Texas, and Florida. In addition, a variation of the process is being used to address wildlife and other water resource concerns in the California Bay–Delta and in south Florida.

Moreover, the Clinton administration has placed substantial emphasis on multiple-objective watershed planning processes as the framework for mitigation banks and other wetland conservation approaches. By 1993, even the Corps was reconsidering its regulatory and construction programs in the context of watershed planning. A report by the Institute of Water Resources, a Corps policy think tank, recommended that "a watershed restoration focus can be used to better integrate the regulatory program with project planning and with the operation and maintenance of existing projects." By linking regulatory decisions to watershed management, said the institute, the emphasis of the program could shift from protecting specific wetlands to restoring wetlands throughout watersheds. Thus, the current trend in the wetlands regulatory program is away from absolute protection, except for high-quality resources, to a planning and management approach that emphasizes multiple objectives, including wetland restoration and mitigation banking.

While NEPA focuses on "major federal actions," it promotes the "scoping" of a spectrum of reasonable alternatives, an analysis of their impacts, and the selection of one or more preferred alternatives, with the understanding that the final determination regarding the alternative to be selected is left to the formal decision makers.

The use of such an issue-focused planning process has a number of benefits. It provides the opportunity to develop a reliable, broadly accepted base of scientific information; to develop the trust that will be necessary to resolve subsequent policy decisions; and to explore policy options before they are lost prematurely as the result of individual actions.

Unfortunately, until relatively recently, there were no assurances provided by the public to private sector economic interests that agreements reached in the plans would be honored. As the plans increasingly became more detailed and represented specific agreements and understandings, the need for assurances increased. Not surprisingly, high-growth states, such as California, Florida, and Hawaii, have provided for development agreements and for implementation agreements in HCPs that would assure, for the benefit of the private participants as well as the agencies, that the plans would be honored.

Placing Mitigation Banking within the Planning Context

The planning processes described earlier provide an institutional framework within which the outcome of a mitigation bank is significantly more predictable. Requirements for success are better understood and regulatory agencies are more likely to cooperate in making a bank effective. In addition, plans may incorporate alternative programs that serve many of the functions of banks. For example, as part of a regional or watershed plan to conserve habitat, a habitat conservancy may be established that obtains funds from a variety of sources (such as impact fees and taxes) and that acquires land and creates, restores, and enhances habitat. The conservancy's costs might be reimbursed or partially offset by payments from developers seeking to mitigate project impacts. This simpler process has the advantage of assuring that conservation has occurred and will be maintained, of accurately quantifying the cost based on actual experience, and of converting the mitigation process to a broader, more proactive conservation initiative. Furthermore, the conservancy's role can be played by one or more entrepreneurial banks.

Because of improved technology, (e.g., computerized mapping and geographical information systems), information regarding economic development and wetlands or wildlife is increasingly available within broad areas such as watersheds and entire ecosystems. This was a key missing ingredient in allowing effective evaluation of mitigation bank proposals.

The changes discussed earlier in the legal/institutional framework should allow mitigation banks to function effectively and, in turn, allow agencies to focus on strategic planning rather than on definitions of specific "in-kind" or geographic proximity criteria. The key question for the agencies is the extent to which a bank's conserved habitat will mesh with the agencies' strategy for maintaining or enhancing biological diversity within a specific watershed or habitat range. To answer that question, the agencies must constantly assess the status of protected resources and those that are at risk. The concept of "gap analysis" has been an important development toward meeting that need. A gap analysis determines the extent of a resource and its assured protection (normally expressed in geographic terms). An agency is then encouraged to focus on significant resource areas at risk.

So where are we now? In the mid-1990s, with a quarter-century of experience and institutional development behind us, the idea of mitigation banking is enjoying a rebirth. There are currently over 50 mitigation banks in operation, with about a dozen states expressly authorizing mitigation banks. In addition, the concept of mitigation banking has expanded to include upland habitats. In April 1995, California promulgated guidelines for "conservation banks" that are intended to broaden the concept of banking to fit within an ecosystem and regionwide conservation planning, such as the NCCP program mentioned previously. Currently, there are a handful of upland habitat-focused banks in southern California, one of which is the Carlsbad Highlands Bank initiated by the Bank of America.

Yet, despite the growing interest in mitigation banking, a number of crucial issues, summarized herein, remain unresolved. Many of these issues are addressed throughout the book.

Assurances The promulgation of formal mitigation banking guidance by federal resource agencies (see federal mitigation banking guidance in the appendix) in November 1995 provides some comfort to potential bankers in that the agencies condone and even support the concept of mitigation banking. Nonetheless, investors still seek greater assurance regarding the available market for credits. Likewise, regulators need assurances that banks will perform as promised and that a bank's wetlands will be preserved in perpetuity. Typically, the Corps receives no such assurances for created or restored wetlands that are not part of a mitigation bank. In fact, most permits are issued with little more than a promise that the work will be carried out. Presumably, mitigation banks will be held to a higher standard. The Corps could require that banks provide a performance bond to guarantee that a created wetland will perform as planned.

Criteria for Success What criteria should be used to determine whether a wetland has been created or restored successfully: Plant cover? Number of birds? Whether the wetland is self-sustaining? For most on-site wetland creation or restoration efforts, the Corps applies simple standards for performance, for example, 80 percent of the site covered with grasses after three years. Typically, after a few years, the permittee may walk away from the project with no further obligations. Will the same standards apply to mitigation banks? Who should be held liable if a created or restored wetland fails?

Timing Under current regulations, development in wetlands may proceed before or while a replacement wetland is created as compensation. Yet, many wetlands never perform as planned. In addition, many wetland values and functions will not be fully replaced or restored until the created wetland is fully established, which could take over 15 years depending on the type of wetland. When, then, can a bank begin selling credits: as a wetland is being constructed, when the work is completed, or when the created wetland is considered successful? Further, how can regulators account for the temporal loss in wetland values: the time between when a wetland is destroyed and its functions and values are fully replaced?

Location It appears as though the standards for wetlands created in a mitigation bank will be more stringent than for wetlands created on-site. Federal regulations, however, currently favor on-site replacement of wetlands over off-site. Under what circumstances would a permit applicant be able to purchase credits from a mitigation bank, rather than create a wetland on-site? Moreover, regulators prefer that created wetlands be located in the same watershed as the destroyed wetland, which means that bank owners must somehow anticipate where future wetland losses will occur in order to establish a bank in the same watershed.

Long-Term Maintenance Over time, many created or restored wetlands require periodic maintenance or remedial work: supplementary planting, weed control, sediment removal, or regrading slopes. How long will mitigation banks have to be maintained? Who will pay for long-term maintenance? Many bank owners plan on donating their wetlands to a nonprofit organization or government agency once all the credits are sold. Should banks be required to establish a fund for long-term monitoring and maintenance?

Currency Evaluating functions and values of wetlands, both those to be filled and those created for a bank remains a sticky issue. How can one measure the value of a wetland? What method should be used? What values are added when a degraded wetland is restored? And, most importantly, how can these values be converted to some form of currency or credits? Some of the factors to be considered when calculating the number of credits created are the value of the wetland destroyed, its location and the location of the bank (same watershed?), whether the functions created are the same as those lost, temporal losses, and the likelihood of success.

Exchange Ratios If a created wetland is equal in value and function to a filled wetland, then the exchange ratio should be 1:1, that is, one acre of a created wetland exchanged for one acre of filled wetland. In December 1991, EPA's Region 9 issued guidance on establishing mitigation banks, which stated that in the absence of information on the specific functions and values of wetlands, use a minimum ratio of 1:1. But, to hedge against the risk of failure, or to account for the difference in values between a filled and a created wetland, the Corps sometimes requires ratios greater than 1:1, say 2:1 or 3:1. The exchange ratio alone could determine whether a bank is profitable or not. For example, requiring a 2:1 ratio instead of a 1:1 ratio could cut the profit of a mitigation bank in half.

Wetland Type Driven by economics, bank owners will likely create wetlands that are easy and cheap to build and maintain. As a result, wetlands that are relatively easy to create, such as marshes, will predominate at the expense of those that are more difficult to create, such as bottomland hardwood swamps and bogs. Should a freshwater marsh be used to compensate for loss of bottomland hardwood swamp? If so, what is the exchange ratio? Is a bog more valuable than a marsh; a mangrove swamp more valuable than bottomland hardwood swamp? Perhaps the exchange ratios should reflect the relative difficulty in creating wetlands: high ratios for bogs and low ratios (1:1 minimum) for marshes.

Sequencing If a wetland in a mitigation bank has successfully been established and is considered fully functional, can a developer skip the first two steps in the sequence of mitigation (avoid, minimize) and simply write a check for the wetland credits? Should the sequence depend on whether a bank is part of a broader conservation plan?

Organization of the Book

The succeeding chapters describe "the basics" of mitigation banking and explore many of the fundamental issues to be resolved in planning and managing banks. In chapter 2, Structure and Experience of Wetland Mitigation Banks, James McElfish Jr. and Sara Nicholas analyze the primary characteristics of the 46 operating mitigation banks identified by an Environmental Law Institute study in 1992. They examine the organization and functions of the banks and describe their track records, including the variety of reasons for failures and successes. They argue that mitigation banking will be effective only in company with other efforts at mitigation, including sequencing and on-site mitigation.

John Studt and Robert Sokolove, in chapter 3, Federal Wetland Mitigation Policies, provide a detailed examination of federal laws affecting mitigation banking. They trace the sometimes conflicting perspectives of the Corps and EPA regarding the appropriateness of mitigation banking and describe requirements codified in the 1989 Memorandum of Agreement between the two agencies that call for a sequence of mitigation policies allowing, as a last resort, use of mitigation banks. Studt and Sokolove also describe provisions of the Regulatory Guidance Letter on mitigation issued by the agencies in 1993.

Chapter 4, State Mitigation Banking Programs: The Florida Experience, by Ann Redmond, Terrie Bates, Frank Bernadino, and Robert Rhodes, describes efforts by the state's Department of Environmental Protection and five regional water management districts to improve the effectiveness of wetlands mitigation. The chapter explores the opportunities presented by new state legislation in 1993 and subsequent rules adopted in 1994 that encourage the use of mitigation banking. The authors detail current experience of state and regional agencies with facilitating the establishment and monitoring of mitigation banks.

Chapter 5, Point/Counterpoint: Two Perspectives on Mitigation Banking, includes an environmentalist's perspective by Jan Goldman-Carter and Grady McCallie, who argue that experience with mitigation banks to date has found them inadequate as a means of replacing natural wetlands. According to Goldman-Carter and McCallie, banks are difficult to build and maintain, and existing methodologies for measuring their effectiveness are unreliable. They suggest that banking should be used only in limited cases and with specific safeguards to limit speculation on mitigation credits, ensure bank performance, and provide perpetual protection.

For the private sector's side of the argument, Virginia Albrecht and Michelle Wenzel argue that mitigation banks would add certainty, predictability, and fairness to the wetlands permitting process while providing incentives for the private sector to enhance, restore, create, and preserve wetlands. Under mitigation banking, an applicant's money would go toward protecting the environment rather than being squandered in a lengthy permitting process. Moreover, according to Albrecht and Wenzel, the environment would be better served by mitigation banks because mitigation obligations would be undertaken by experts whose professional reputations depend on the outcome of their work.

Leonard Shabman, Paul Scodari, and Dennis King tackle the difficult issue of establishing private credit markets for mitigation banks in chapter 6, Wetland Mitigation Banking Markets. They define and discuss the conditions necessary for the widespread emergence and ecologic success of private credit markets for mitigation banks and explore the interaction of credit suppliers, permit applicants, and regulators in the operation of credit markets. They argue that the best course to advancing the private credit market is to set trading rules according to environmental criteria and then to allow the applicant and supplier bargain over credit prices without interference from public agencies.

In chapter 7, Legal Considerations, Lindell Marsh, Robert Sokolove, and Robert Rhodes examine the various legal issues and requirements raised in establishing mitigation banks. They identify six types of enabling instruments used to establish banks, set out the terms of a banking agreement, and describe procedural considerations. In addition, the authors evaluate the usefulness of an impact fee alternative to mitigation banking.

John Rogers, in chapter 8, Wetland Mitigation Banking and Watershed Planning, examines the value of considering mitigation banking as part of a broader watershed management strategy. He suggests that in planning for watershed management, suitable sites for effective mitigation banks can be identified and wetlands can be prioritized for preservation through advanced identification, thus allowing more flexible application of mitigation rules throughout the watershed.

Chapter 9, The Practice of Mitigation Banking, by Lindell Marsh and Jora Young, describes the multitude of concerns that must be addressed in planning to establish a bank and in managing an established bank, including defining the approach, assembling a team, selecting a site, conducting a feasibility study, and establishing a maintenance program.

In the Conclusion, the editors highlight the approaches tried and lessons learned from experience to date and suggest some issues that still remain to be resolved if mitigation banking is to flourish and achieve conservation objectives.

Mitigation banks come in all shapes, sizes, and institutional arrangements: single-user banks, public agency banks, and entrepreneurial banks. Following the conclusion, eight case studies provide brief profiles of a variety of mitigation banks that are underway in different parts of the country—from entrepreneurial banks in Georgia and Florida to multiagency banks in California. The case studies show how mitigation banks can be used to offset losses to wetlands or wildlife habitat.

2

Structure and Experience of Wetland Mitigation Banks

James M. McElfish Jr. and Sara Nicholas

Wetland mitigation banking is a relatively recent phenomenon.[1] In 1983, a Corps of Engineers study of wetland mitigation banking revealed that while there were several proposals for such banks, none was operating yet.[2] The Tenneco LaTerre Bank (subsequently called Fina LaTerre) in Terrebonne Parish, Louisiana, was just gaining its approvals; its Memorandum of Agreement (MOA) was signed in January 1984. A comprehensive study by the U.S. Fish and Wildlife Service in 1988 identified 13 banks, all commenced during the 1980s.[3] Since 1980, a number of states have enacted legislation authorizing the establishment of wetland mitigation banks, with most of this legislation enacted after 1985.

In 1992–93, the Environmental Law Institute (ELI) studied all operating (and many proposed) wetland mitigation banks in the United States to assess their status and to determine what approaches might have value for future banks. This study was supported by the Corps of Engineers and the Environmental Protection Agency.[4] ELI identified a total of 46 existing banks as of mid-1992. An existing bank was defined as either (1) having a signed memorandum of understanding or similar instrument (e.g., permit), rendering it "open for business" or (2) having already issued credits with the acquiescence of one or more regulatory agencies.

The 46 banks were located in 17 states. Eleven were in California, which recognizes mitigation banking specifically in state law. Eight were in Florida, all but one in the Southwest Florida Water Management District (SWFWMD). California and Florida led in the number of existing wetland mitigation banks, primarily because of strong development pressures in both states throughout the 1980s and because state and local regulators were willing to experiment with the concept. No other state had more than four existing banks (although the Minnesota Highway Bank had over 40 separate mitigation sites).

Most of the banks (42 of the 46) served only one user. Nearly 75 percent of all banks were state highway banks, port authority banks, or local government banks providing mitigation solely for public works projects. Indeed, 22 were operated by state departments of transportation to mitigate for highway construction. (In 1993, an additional transportation wetland mitigation bank was authorized by Maryland.) Six banks were controlled by private developers and used solely for advance mitigation of their own proposed projects.

In 1992, only one privately owned mitigation bank offered credits for sale to the general public: the Fina LaTerre Bank (which, nevertheless, used the majority of credits for itself). There were only three publicly owned, or nonprofit-agency-owned, wetland mitigation banks offering credits for sale to the public—the Bracut Marsh Bank and the Mission Viejo/Aliso Creek Wildlife Habitat Enhancement Project (ACWHEP) Bank in California and the Astoria Airport Bank in Oregon.

However, as experience with banking has grown, so has the development of commercial wetland banking. In 1993 two more commercial wetland mitigation banks were opened—the Millhaven or Wetlands Environmental Technology (W.E.T., inc.) bank in Georgia; and the Florida Wetlandsbank, also known as the Pembroke Pines bank, in Broward County, Florida. Another commercial mitigation bank, the Ohio Homebuilders' Bank, was approved and opened in 1994. Of over 60 proposed banks ELI identified in 1992, more than half were commercial wetland banks.

Bank Organization

Mitigation banking arises from the need to reconcile two competing sets of interests: those of the private developers or government development agencies whose activities will have some impact on existing wetlands that are protected by law and those of the government agencies charged with protecting wetlands. Thus, every banking program has at least two parties. But many are far more complex.

As evidenced by the number of mitigation banking agreements between state departments of transportation (DOTs) and the Corps of Engineers and state permitting agencies, it is entirely possible for developers and agencies to establish successful banks purely as an offshoot of the existing permitting process, without complex governance structures or other parties. However, even in the simplest DOT bank, the two parties perform several discrete functions, which, in more elaborate versions of mitigation banking, often are separated from the permitting process and delegated to a number of other entities.

Functions

While mitigation banks vary widely in structure, every banking program includes six functions: (1) client (use of credits), (2) permitting, (3) credit production, (4) long-term property ownership, (5) credit evaluation, and (6) bank management.

The apparent diversity among banks largely results from the different ways in which these six functions are allocated among the various parties. As just noted, in the simplest banks, the functions are divided (or shared) between a developer and the permitting agency or agencies. As mitigation banks become more complex and the division of labor in the area of wetlands mitigation becomes more specialized, these same six functions could be performed by as many as six different parties, some of which have no direct connection to the permitting process.

1. The bank's reason for being depends upon the existence of a client, the entity or entities whose activities will create a wetlands impact for which mitigation is being sought through the bank. A client thus is identical with a would-be permit holder and can be any private or public development entity whose project triggers the permit requirements. While these entities typically play several roles in the banking process, in their role as clients they represent market demand for compensatory mitigation credits and need not have any involvement in the actual mitigation work or possess any attribute other than a sheer willingness to pay for the mitigation credits.

2. The permitting function involves deciding whether a project affecting wetlands, and for which mitigation may be required, will be allowed to proceed. It generally is exercised by the government agencies, federal, state, or local, with jurisdiction over affected wetlands. Often, there are several such agencies with concurrent jurisdiction and varying degrees of oversight; representatives from each agency sometimes form an interagency committee that makes permitting decisions. In the case of wetlands regulated under Section 404 of the Clean

Water Act, agency responsibilities can range from commenting (U.S. Fish and Wildlife Service and other federal and state resource agencies) through permit writing (the Corps of Engineers and state water control agencies) to veto power (EPA). By establishing requirements, such as sequencing or proximity restrictions, that determine whether banking will be an acceptable form of mitigation in specific cases, the permitting agencies effectively create the market for mitigation banking and exert substantial control over the regulatory climate in which banking will occur.

3. A third essential function is creation of mitigation credits, the physical wetlands commodity whose value is traded or sold by the bank. The credit production function entails the production of viable wetlands credits on a specific mitigation site or sites by any of the accepted methods: restoration, creation, enhancement, and, in certain cases, preservation. In more concrete terms, the credit producer generally is the chief proponent of the plan for creating credits, acquires initial title or other right of entry to the site, and carries out the mitigation work. While some of these tasks may be contracted out or otherwise delegated to an agent, the credit producer bears primary financial and legal liability for successful construction and development of the mitigation site and often for subsequent monitoring and maintenance as well.

Credit production can be performed by the client or by the permitting agencies, but it remains analytically distinct from either function. It is also possible for a third party, such as another government agency, a private entrepreneur, or a nonprofit organization, to produce and sell mitigation credits acceptable to both of the parties to the permitting process. Indeed, such third parties may become more proficient at acquiring suitable mitigation sites and producing surplus credits than either permitting agencies or full-time builders of highways and condominiums.

4. Given the desirability of creating enforceable legal mechanisms that will ensure the mitigation site is maintained as a wetland for an ecologically useful period of time, it is important to identify and isolate the function of long-term property ownership. While, as noted, the credit producer often holds fee title, a conservation easement or other right of entry to the mitigation site, ownership of these rights is a separate function that can be transferred to or exercised by parties not otherwise involved in the banking process.

For example, it is fairly common already for credit producers to transfer their property rights to state or federal natural resource agencies or nonprofit groups like The Nature Conservancy, either during the bank's life or after all credits have been used. Conversely, entrepreneurs that hold a large quantity of land with potential for wetlands creation or enhancement could elect to retain their property rights while allowing "mitigation farming," where credit producers pay for the right to create credits on a specific parcel without assuming ownership. In each of these cases, the primary function of the long-term property owner is to exclude any other uses of the land that would interfere with its continued existence as a dedicated wetland.

Depending on the precise nature of the property right being held, long-term property ownership may entail other duties assigned by applicable property or contract law. These could include active monitoring and maintenance of the wetland and financial liability for remedying mitigation failure or any damage to third parties—responsibilities that generally fall to the credit producer but also can be assigned contractually or as a condition on transfer of property rights.

5. Once the wetlands credits have been produced, both they and the impacts they will mitigate must be quantified to conform to the currency in which the bank is trading. Since credit producers have a financial stake in maximizing the value of credits and clients have one in minimizing the value of impacts, credit evaluation often is done by one of the permitting agencies or by an outside party, such as another resource agency or independent consultant acting as a wetlands "appraiser." Even in banks where the field work underlying credit evaluation is performed by a credit producer or client, final review of this work by one of the permitting agencies is standard, thus ensuring some independence for the credit evaluation function.

For instance, the proposed entrepreneurial Springtown Natural Communities Reserve bank in Livermore, California, would delegate the credit evaluation function to the California Department of Fish and Game to avoid any potential conflict of interest. Similarly, a model memorandum of understanding drafted by the Federal Highway Administration to assist state DOTs in their banking efforts calls for the creation of a "Technical Subcommittee," which is composed of one member each from the state DOT, the state department of fish and wildlife, and the local office of the Corps of Engineers. This technical subcommittee is charged with assessing proposed impacts using any "appropriate methodologies," including the Habitat Evaluation Procedures (HEP), Wetland Evaluation Technique (WET), or best professional judgment. Versions of these evaluation procedures have been adopted by DOT banks in Arkansas, Montana, and Nebraska.

6. Bank management is the process of determining whether produced credits and proposed debiting projects meet the conditions established for use of the mitigation bank and recording resulting transactions. In single-client banks, like many of the DOT banks, this function is minimal and largely inseparable from the permitting process itself: the client and the permitting agencies agree in advance on the bank site or sites, subsequent use of which is reflected in an informal ledger kept by one of these parties, which records each "withdrawal" of credits and updates the balance accordingly.

In more complex schemes where several different parties are producing credits and several others are purchasing them, the bank management function may be delegated to a wholly or partially independent individual, board, or trust charged with the fiscal management of funds and banked credits in accordance with criteria that may differ from those considered in the permitting process. For instance, in the proposed wetland banking system for Prince George's County, Maryland, the county government would name a bank manager who would have approval power over proposed debits of one acre or less and inform the clients and permitting agencies whether mitigation is available from the bank.

For larger debits, the bank manager would make a recommendation on the propriety of bank use to an interagency oversight team. This two-tiered process relieves the permitting agencies from having to review the bank's status each time a routine debit is proposed. Similarly, the Minnesota Department of Transportation bank uses a team of bank managers, with representatives from the DOT and each of the various permitting agencies, to decide which projects will be accepted for bank debits or credits.

Banking's Track Record

The jury is still out on whether wetlands mitigation banking, in general, and individual banks, in particular, have achieved "success" in replacing wetland values. Most active banks have been in existence for fewer than ten years, and the time to full functional replacement of wetlands values—while it will vary considerably by wetland type and region—is often much longer. Like on-site mitigation projects, a number of wetland banks have failed.

In general, mitigation projects fail for two reasons. First, the project may be improperly sited, designed, or constructed. Second, a functioning project may be damaged by subsequent events. Both of these causes of failure require attention at the outset of a banking scheme.

The most common failure is improper design or construction of the mitigation site's hydrology. If a site's elevations are incorrectly surveyed or constructed, for example, few of the anticipated wetland functions will be realized. In one study of on-site mitigation in the South Florida Water Management District, researchers determined that 25 percent of all mitigation projects were suffering from "significant" hydrologic problems.[5] One of the first mitigation banks approved by the Southwest Florida Water Management District, the Northlakes Park Bank in Hillsborough County, Florida, failed because of improper hydrologic design. Although credits were recognized and debited, the bank failed in its attempt to rehydrate a forested wetland. The Fort Lee Mitigation Bank in Virginia also failed to achieve the expected hydrology. Although this is a common problem, some types of wetlands are more susceptible than others to this kind of failure. Emergent wetlands surrounding open water require less precision than, say, forested wetlands or estuarine marshes.

Other common failures include failure to identify existing problems with substrates, soils, and contaminants. The Bracut Marsh Bank in Humboldt Bay, California, for example, was sited partly on compacted soils unable to support suitable vegetation, partly on woody debris that formed an unstable substrate leading to formation of sinkholes in the marsh, and to migration of the debris with tidal flow. The Otterdam Mitigation Bank in Virginia had higher than expected construction costs because of a clay layer at the site that was not identified when the site was selected. A nonbank wetland restoration project sited at Sweetwater Marsh in San Diego Bay encountered both a hazardous waste landfill and a construction debris landfill, both of which had to be excavated and removed for disposal at substantial expense.[6]

Site selection difficulties can also arise from failure to consider surrounding land uses that may impair the long-term viability of the mitigation site. Mitigation sites without upland buffers or that are surrounded by impervious surfaces can quickly convert to uplands or become pollution sinks. The Batiquitos Lagoon bank in southern California is subject to heavy siltation from adjacent uses.

Sometimes the site requires more active or continuous manipulation than is practicable. The Mud Lake bank in Idaho failed because designers did not anticipate the difficulties with keeping the site hydrated. The selected design called for continuous pumping of water onto the site. Unfortunately, insufficient water was available for mitigation because of competing irrigation and development uses, exacerbated by drought conditions. Moreover, the water that was pumped to the site rapidly leaked through cracks in the hardpan that formed. The pump then fouled and failed to operate. The bank, which was established in 1990 to mitigate the loss of 16 acres of wetlands, failed to do so; the site was completely dewatered and the vegetation was predominantly upland.

Poor plant selection, failure to sustain plants during the establishment phase, and improper planting depths are also common start-up failures. So is the failure to import a growth medium where the on-site soils are inappropriate for plant establishment. Even if sites are properly selected and well designed, some initial failure with revegetation can be expected and should be planned for. Vegetation may not do well initially for a variety of reasons. For example, at Pridgen Flats, a pocosin restoration bank in North Carolina, an adequate number of growing plants could not be obtained, so seeds were used for much of the site. However, none of the seeds (sowed in spring 1992) germinated.

Construction-related accidents also cause problems. During preparation of the site for the 4.2-acre Naval Amphibious Base Eelgrass Bank in San Diego, California, for example, the Navy's dredging activities accidentally destroyed 6.2 acres of natural eelgrass. Fortunately, the damaged area recovered on its own three years later. The bank site itself was less successful; initially, only 1.6 acres achieved successful vegetation. Similarly, the (nonbank) Sweetwater Marsh restoration project had its plant nursery accidentally bulldozed by contractors working on part of the mitigation. The (nonbank) Irvine Company wetland restoration project near the University of California at Irvine had its vegetation killed in successive years, first by misapplication of herbicide and then by failure of the irrigation system.

Other common problems facing mitigation sites include vandalism, natural disasters (e.g., storms, fires, floods), ice damage, off-site activities (oil spills, damage from powerboat wakes, loss of storm protection from barrier islands), accumulation of debris, and invasion by undesirable exotic species, diseases, and insect pests. A bank can also be a victim of its own success. For example, a mitigation site may be so attractive to wildlife that all the vegetation is eaten and the site is left vulnerable to erosion or washout of the substrate. Many of these failures are not preventable, but are (at least in the aggregate) predictable. The credit producer's, or long-term landholder's, responses to foreseeable failures should be planned in advance.

Preventing Failures: Performance and Design Standards

Most failures can be avoided at the design and construction phase using performance standards or design standards. In its simplest form, the performance standard approach simply requires the agencies responsible for recognizing the bank credits (or allowing their use) not to allow use of the credits until the project is fully functioning. This is consistent with the notion of mitigation banks as providing advance mitigation for development activities. It is simple and effective, although it is not necessarily easy to define when a project is fully functional.

However, advance mitigation is not the rule among current and proposed banks: many banks and banking schemes allow the use of credits prior to full success of the bank. These schemes simply hold the credit producer liable for correcting problems in the event of failure. Some also require a greater compensation acreage ratio for use of credits prior to full functional replacement.

Banks that allow use of credits prior to their full functioning cannot, however, reasonably rely on performance standards alone. While, in theory, enforcement would assure the prompt correction of any failures, enforcement does not always occur. Even when it does, it is not always effective—especially where the development activity has already been completed. In such cases, the regulatory agency's leverage to obtain corrective action is diminished because the developer has already realized the benefit and has no incentive for rapid compliance; and the credit producer (if a different entity) has already been compensated for the credits and has no incentive for rapid compliance. The Northlakes Park mitigation bank in southwest Florida, for example, has not produced its wetland credits even though they were all expended immediately upon regulatory approval of the bank; the bank is in debit status. This is the same problem that has afflicted on-site mitigation.

A more prescriptive approach is the use of design standards. Typically, this requires submitting site assessments, plans, and detailed construction and operating information to a regulatory authority before receiving approval to generate credits at a mitigation site. The regulatory agency requires sufficient information to assure itself that the mitigation project is likely to succeed.

A number of existing banks submitted detailed design information as a condition of their approval. Others defer this step until after the bank has been approved. The draft MOA for the proposed Neabsco (Virginia) Bank, for example, states that initial designs must be submitted and that final designs are to be agreed to by mutual agreement between the credit producer and the Corps. Oddly, however, this draft proposes that in the event of a difference of opinion on design the matter is to be submitted to arbitration. Normally, the permitting agencies retain the final say on design decisions where submissions are required. Reliance on design standards may impose additional costs and reduce credit producers' flexibility. However, where a regulatory agency has reason for concern about performance (either because it is allowing some drawdown of credits prior to bank success or because there are few or unique potential mitigation sites), design standards provide a rational approach.

Design standards may include requirements for preliminary site assessments, proposed design parameters, timing of construction activities and identification of materials, substrate, growth medium, and vegetation. They may also require certification of designs by persons with relevant professional training, monitoring of the construction activities, submission of as-built drawings and progress reports, and other information. The Mission Viejo/ACWHEP Bank agreement not only provided for agency review and approval of designs but required a $10,000 payment from the credit producer to the county government to fund a consultant to monitor the bank's adherence to those standards.

Even projects subject to design standards and quality control requirements can fail. For example, the Bracut Marsh Bank was constructed to exacting engineering specifications. Unfortunately, the specifications proved to be incorrect to allow regular tidal flushing of the bank site. Even after this was first discovered—six years into the project—necessary changes were not made. Because of such instances, design standards should be backed by performance standards. The Weisenfeld Bank in Florida, for example, has detailed design specifications backed by success criteria. Some consultants have suggested that where a design has been approved and accurately carried out, if it is unsuccessful, the mitigation should be deemed complete. However, the regulatory objective is to accomplish functional replacement, not just expenditure of good faith effort. Design standards are not a substitute for success but a further guarantee.

Quality control can also be effective in preventing unnecessary failures. Mitigation projects designed by competent engineers, biologists, and other experts are more likely to succeed. An accreditation process could be adopted to distinguish the qualified from the unqualified. The accredited restoration expert would then certify the design and construction of the project. The qualifications cannot simply be professional degrees, however. Because wetland restoration is a fairly new field, many of its qualified practitioners have degrees in the "wrong" fields, or no degrees relevant to an area of work that they have learned primarily by practice. If there is accreditation of persons or firms, it should be based on objective measures, such as examination and/or experience requirements.

Although no existing bank requires accreditation or certification, this is not an unusual way of preventing siting and design failure. For example, virtually all state and federal regulatory programs for reclamation of mining sites require that the reclamation design be certified by a registered professional engineer. Likewise, building codes require that designs and as-built drawings be certified by trained professionals. As a matter of public policy, these laws do not simply rely on design or performance standards, or the threat of liability in the event of a failure. Rather, the laws are designed to assure that a technically trained person is planning the project and overseeing it to prevent a failure.

Contingency Plans, Maintenance, and Monitoring

Mitigation bank failures occur for a number of reasons, including disease, weather, third-party damage, accidents, catastrophic events, consumption of the wetland vegetation by wildlife, and others. Banking schemes must anticipate these events. Many do not.

The only rational approach to such sources of failure is to plan for them. The most foreseeable failures are those from natural occurrences. For example, the occurrence of a 10-year storm event, even a 100-year storm event, can be expected during the life of a bank. If the bank has not been designed for these events, it is operating in a fantasyland. It is reasonable, therefore, for the regulatory agency to insist on knowing what the bank operator intends to do should one or more of these foreseeable ills appear. The advantage of requiring a contingency plan is that it compels the operator to consider these factors at the planning stage and to ascertain what, if any, preventive (as well as remedial) measures can be taken.

The Florida Department of Environmental Regulation, in its wetland mitigation banking policy, requires that mitigation banks have contingency plans. These plans must be updated semiannually. Again, there are useful analogies to mining reclamation. California law, for example, requires mine operators to prepare contingency plans that identify what the operation's response will be if catastrophic events or maintenance failures occur. Plans must be updated periodically as conditions change.

Maintenance can be important in preventing failures. The appropriate level may vary significantly based on the type of bank. The Fina LaTerre bank requires substantial maintenance, because it depends on active management in order to generate credits from the avoidance of saltwater intrusion. Others, such as the Company Swamp bank, require little maintenance because they involve the preservation of stable ecosystems—in this case an old-growth cypress swamp.

A formal monitoring system is an important element of wetland mitigation banking. Sufficient monitoring should occur to detect as quickly as possible instances where the site has been denuded of vegetation by muskrats or damaged by ORV (off-road vehicles) enthusiasts. One of the major failings of Bracut Marsh was that no monitoring was done for the first six years. This made it difficult to take meaningful corrective action. If there is a reasonable monitoring program, coupled with a set of performance standards, and a contingency plan for future failures, the banking instrument may not need to specify a particular maintenance program.

The question of who does the monitoring is an important one. One option is to require the credit producer to monitor the bank. This is the approach taken in many of the early MOAs and in the more recent Section 404–permitted banks. Requiring self-monitoring of a single-client bank, such as a DOT bank, forces the party causing an environmental impact to be responsible for making sure that its mitigation efforts are successful. Self-monitoring can also make enforcement easier because the enforcing agency uses the permittee's own data to prove noncompliance.

The Weisenfeld Mitigation Bank in Florida is a good example of how self-monitoring can also be used to promote self-enforcement. Weisenfeld was granted a banking permit from the Florida Department of Environmental Regulation, which not only requires Weisenfeld to send regular monitoring reports to the department and several other agencies but requires Weisenfeld to police itself. If the bank discovers that it is not in compliance with the permit conditions, Weisenfeld must immediately explain to the department the type and the cause of noncompliance as well as the expected duration of continuing noncompliance and the steps being taken to return to compliance. The department expressly reserves the right to inspect, sample, and monitor the bank site. The permit also clearly states that all records and monitoring data submitted to the department may be used as evidence in enforcement cases.

In a multiclient bank, self-monitoring may be more complicated. Self-monitoring can be the responsibility of the credit producer, landowner, or the clients. The proposed Homebuilders Association of Greater Chicago Mitigation Bank MOA provides detailed success criteria and specifies that corrective measures must be undertaken by the credit producer. During the construction phase, inspections by "qualified individuals" (presumably employed by the bank) must occur no less than monthly (and within one week of any rain event). The results of the construction inspections must be submitted to the Corps. Then "intensive monitoring" must occur for not less than five years from the date of credit production or three years from the last sale of credits, whichever is later. "Limited monitoring" then occurs every other year for 15 years from the end of the intensive monitoring period.

Typically, existing mitigation banks are monitored on an ad hoc basis by the regulatory agencies permitting the development activities. The responsibility of monitoring can also belong to an interagency team, usually made up of the signatories of a bank's MOA. It is preferable to spell out monitoring obligations in more detail to assure that they are fulfilled.

Another option is to have a state or federal inspector paid for by the credit producer. The Mission Viejo Company, for example, provides funding to Orange County for inspection of the Aliso Creek Wildlife Habitat Enhancement Project. Bank funding of inspectors is analogous to New York's statutory requirement that commercial operators of hazardous-waste treatment, storage, and disposal facilities fund an on-site inspector employed by the state's Department of Environmental Conservation. At least one proposed bank (Prince George's County, Maryland) will assign a bank manager who is responsible for inspecting the bank at least annually, and after all major storms, for at least five years. The manager will submit all monitoring reports to an interagency oversight team. This system is appealing because it makes one disinterested individual accountable for monitoring the bank.

A designated bank monitor could easily be required to have professional qualifications—for example, biologist, hydrologist, engineer. Requiring certification of bank monitors is one way to try to assure accurate evaluations. Another is to require monitoring reports to be signed and certified by responsible company officials, as is the case with the discharge monitoring reports (DMRs) required under the Clean Water Act. Having certified reports should minimize the falsification of data by credit producers. The Weisenfeld Mitigation Bank in Florida makes it a condition of the banking permit that all monitoring information must include the name of the person responsible for performing sampling, measurements, and analysis.

At least as important as who does a bank's monitoring is who gets the results of those evaluations and what they do with the data. Monitoring results can go to the permitting agency, interagency oversight teams, and/or the public. The reports can be used to evaluate or reevaluate credits, to determine whether or not performance standards are being met, and to demonstrate compliance with permit conditions.

Assigning the Risk of Failure

Critical to any banking scheme is clarity regarding responsibility for correcting any failures. Banking programs that assume flawless performance by all participants (and perfect cooperation by nature) are too common. The issue arises most often where a bank has been allowed to sell credits prior to achieving full performance. A somewhat different problem arises in instances where a bank site fails after all the credits have been issued or debited. Who should bear the risk if the mitigation fails? The assignment of liability should be explicit in any banking scheme rather than implicit or unspecified. Among the options available are the credit producer, the client(s), the site owner, no one, or the regulatory agency.

The most obvious candidate for responsibility is the credit producer. The credit producer undertook to provide the credits and should have planned for contingencies. The regulatory agency and the clients both relied on the credit producer to produce the wetland values and functions now damaged or destroyed. Presumably the credit producer also has the expertise, site access, and resources to take corrective action—and was paid for the credits.

Alternatively, the clients might be liable. They would have been liable had the mitigation been on-site. They are the ones who benefited from the use of the banked credits. Arguably, if the credits turn out to be no good (or less valuable than represented), the clients should make good on them. The difficulties, however, are that the clients may well lack access to the site (they probably relied on the bank in order to avoid long-term ownership, liability, and maintenance responsibilities) and have trouble untangling responsibility for corrective action at a multiclient bank. (Each client is not necessarily responsible for an identified parcel but rather for a set of credits generated as a whole.) This makes assignment of responsibility difficult.

Another alternative is to hold no one liable. Natural wetlands experience losses all the time. It may be irrational to expect more from created, restored, or enhanced wetlands than from the wetlands they replaced. A possible approach is to hold no one liable where the event is one that would have (could have) destroyed the wetland for which the mitigation was required—for example, a 100-year storm event, a hurricane, a regional infestation—but to require rehabilitation in all other instances.

A related issue is what to do with constructed but unsold credits that are destroyed. Probably the best approach is not to recognize them as available unless they (and the rest of the bank) are rehabilitated. Thus, the credit producer bears the risk of loss for any unsold credits. However, the interagency guidelines for wetland mitigation banking prepared by the Corps of Engineers' Galveston District make no one responsible for failed or destroyed credits. The guidelines provide that "once the credits have been established, they will remain until all of the credits have been withdrawn. Credits will not be adjusted up or down . . . even if the mitigation bank exceeds expectations or does not meet expectations."

Another possible liability scheme is to make failures the responsibility of the long-term owner of the property. The purpose of having a long-term owner is to provide some assurance of the long-term viability of the wetland; responsibility for maintenance and reconstruction may be a long-term adjunct of this responsibility. One difficulty with this allocation is insisting on rehabilitation if the landowner has no direct relationship to the regulatory agency. For example, can a regulatory agency compel a state parks agency or a nonprofit conservancy to take action? This would need to be explicit in the authorizing instrument. The funding issue may be particularly acute here. If the land-owning entity is not the entity that received funds for the sale of credits, it may be necessary to assure that it has a source of funds sufficient to deal with contingencies.

Finally, the regulatory agency itself may assume the liability. Essentially, once it has recognized credits in mitigation of a permitted activity, the agency may release the other parties from liability. This approach is simple and direct; however, it provides virtually no assurance that rehabilitation of a damaged mitigation bank site will take place. Most regulatory agencies do not have the budget, technical expertise, or staff to undertake an active rehabilitation effort.

Existing banking schemes are not extremely helpful in specifying liability. The 22 existing state Department of Transportation banks (in 14 states) use several of these approaches. In Minnesota and North Carolina, the Department of Transportation (client and credit producer) remains liable no matter who ultimately gets title to the land. In New Mexico, the DOT is liable for 25 years; and in Wisconsin DOT is liable until the bank site is deemed "successful." Thereafter, no liability remains. In Idaho and Tennessee, the liability shifts to the ultimate landowner (usually a resource agency). In the Louisiana DOT bank, the liability was unstated and remains in dispute between the DOT and the resource agency landowner. The still-proposed Nebraska DOT banking program leaves liability open for negotiation upon disposition of the site, a difficult time to resolve the issue.

Most non-DOT banks are silent as to liability. However, some specifically make the client-credit producer (the same entity in most existing banks) liable. The draft northeast interagency regional guidelines provide that in the event that a bank fails to provide the compensation required, "the permittee remains responsible for compensating for the wetland functions lost as a result of permitted activity."

Financial Guarantees

Given the possibilities for mitigation failure and the risks in allocating liability, financial guarantees can serve an important function for mitigation banks. There are numerous financial instruments that can serve to guarantee mitigation success, and to provide a source of funds in the event of contingencies.

The best of these guarantees serve dual purposes: (1) to ensure that funds will be available to repair and maintain the site in the event of a problem not corrected by the credit producer, and (2) to provide the credit producer with an incentive to design, construct, and maintain the site properly.

Despite their utility, very few existing or proposed banks have any provision for financial assurance. None of the state DOT banks does. The rationales for not requiring financial assurance from governmental entities are the assumptions that they will always be around to honor their obligations, that as governmental agencies they are likely to do so without resistance, and that to require a financial assurance is either to incur unnecessary government expense (for a third-party bond) or unnecessarily to idle limited government resources (in a fund). The Federal Highway Administration's 1992 draft model banking agreement does not specify financial assurance. Several of the existing and proposed private banks and publicly operated banks do provide for financial assurance, although these are the exception rather than the rule. Like most on-site mitigation projects, most existing banks do not have such assurances.

Some mitigation banking policies and guidance documents require financial assurance, while others do not. The U.S. Fish and Wildlife Service's 1983 interim guidance, while not expressly requiring financial assurance, states that "means for long-term operation and maintenance shall be agreed upon." In contrast, EPA Region IX's 1992 final guidance document for mitigation banking specifies that "a fund for remedial actions should be established as part of any banking agreement." EPA Region V's draft guidance document does not specify financial assurance; while EPA Region IV's draft guidance for mitigation banks provides that financial assurance "should be established as part of any banking agreement" and should be in such form as to "provide an irrevocable guarantee of availability of the necessary financial resources" to cover "bank needs, including but not limited to remedial actions." More recently, the Federal Guidance on mitigation banks, proposed in March 1995, stated that "the bank sponsor is responsible for securing sufficient funds to cover contingency actions in the event of bank default or failure. In addition, the bank sponsor is responsible for securing adequate funding to monitor and maintain the bank throughout its operational life, as well as beyond the operational life if not self-sustaining."[7] According to the Guidance, financial assurance requirements should reflect realistic cost estimates for monitoring, long-term maintenance, contingencies, and remedial actions.

Financial assurance can be provided in a variety of forms: surety bonds, trust funds, escrow accounts, sinking funds, insurance, self-bonds, and corporate guarantees.

The surety bond is the classic approach to assuring performance and preparing for contingencies. The credit producer purchases a bond from a third-party surety (paying a premium and posting collateral) or provides a cash bond, letter of credit, or other assets that ensure that the site functions properly for the specified period and that all necessary corrective actions will be taken. Once the period has ended and performance has been successful, the bond is released. (This may also be done in stages. As certain milestones are reached, portions of the bond are released.) The bond provides both a source of funds that can be drawn on by the regulatory agency (or bank manager, if appropriate) in the event of a default by the credit producer and an incentive for the credit producer to do things right so that the bond can be released.

The Mission Viejo/ACWHEP bank has an $800,000 bond posted by the client/credit producer with the county to assure that construction and vegetation is carried out. The bond is releasable incrementally over five years based on attainment of vegetation milestones. The Millhaven (GA) bank approved by the Corps in December 1992 must post a performance bond of $5,000 per acre. The bond is reduced to $1,000 per acre upon the Corps's verification that the wetland acres are performing; thereafter, the reduced bond remains in effect until the completion of a five-year maintenance period. Each distinct "block" of wetland acres in the bank is bonded separately. The evaluations and reductions are done by "block" in order to avoid the expense and difficulty of bonding (and calculating the bond reductions) for the entire site at one time.

A second approach is the establishment of a trust fund. Unlike the bond, a trust fund is primarily aimed at providing sufficient funds for maintenance and contingencies, not at providing an incentive to the credit producer or bank manager. The Batiquitos Lagoon mitigation bank provided for a trust fund. (The fund was not created as the client—Pac/Tex—withdrew.) The client was to have provided a $15-million initial contribution, which was to generate construction, operating, and maintenance funds for the first 30 years. Concurrently, there was to be a separate fund to earn and reinvest interest so that at the end of 30 years the interest on the accrued balance could thereafter generate annual maintenance funds. The Mission Viejo/ACWHEP bank has a client-created $143,000 trust fund that is intended to generate $10,000 per year for operating and maintenance expenses for a 15-year maintenance period. The proposed Springtown mitigation bank, also in California, has proposed a trust fund financed by a surcharge on the sale of credits. Although the trust fund amounts have not been determined, the bank's proponent suggests that $5,000–$10,000 per acre might be an appropriate amount and that ultimately the fund would generate $60,000–$100,000 per year for operating and maintenance expenses.

The Huntington Beach Mitigation Bank has a "trust fund." The fund is not tied to particular acreage or success criteria and it is not limited to particular expenditures. The fund is produced primarily through contributions and other funds going to the local nonprofit conservancy that administers the bank and has a balance of between $5,000 and $10,000 for the entire site.

Other approaches include escrow accounts and sinking funds. These combine the trust fund approach with incentives to perform maintenance and other required activities. For example, the proposed Chicago Homebuilders bank would have an escrow account. Upon sale of credits at market price, the credit producer would deposit funds into an escrow account "to ensure the long-term monitoring, management, and maintenance" of the bank. The deposit would be $5,000 for each "fully certified" acre sold (credits that have been recognized as successful). For "conditionally certified" credits, the credit producer would deposit $10,000 per acre. This amount would then be reduced to $5,000 per acre once the acres achieved full certification (i.e., a refund). The proposed memorandum of agreement would also allow the sale of credits by a bank that has no assets (viz., that has not yet produced credits or conditional credits). These future credits—sold at market prices and usable at a proposed compensation ratio of 1.5:1—would require the deposit of $30,000 per acre into the escrow account. The credit producer would, however, be permitted to withdraw up to $20,000 per acre to construct the bank credits; and the further reduction of the escrow amount to $5,000 per acre would occur when the credits received full certification. Interest on the escrow account would be usable by the bank for monitoring, management, and maintenance. If the bank becomes insolvent, its assets, including the escrow account, would become the property of the Corps of Engineers. The draft agreement does not address the ability of the Corps to use the escrow account in the event of a dispute with the bank or upon particular defaults, nor does it specify what happens to the principal after the conclusion of the 15-year monitoring period that follows the initial "intensive" monitoring period.

Sinking funds are accounts in which the fund balance is allowed to decline over time as the likelihood of failure diminishes. They can be tied to particular success criteria (vegetation diversity and distribution, for example).

Insurance is conceivably an alternative approach. While insurance may not be commercially available to guarantee a credit producer's banking success, it may be possible to purchase insurance against operator accidents, vandalism, and floods, fires, and storms. Insurance is good at dealing with contingencies; unlike surety bonds, trust funds, and sinking funds, however, it does not have a significant incentive function.

A variation on the surety and insurance approaches is the bond pool. While not currently in use in wetland mitigation, they are common features of state laws concerning coal mining reclamation. Bond pools are risk-sharing mechanisms. The participant in the bond pool posts a site-specific bond in an amount substantially less than that needed to cover all contingency costs, and in addition pays a periodic nonrefundable amount (or a one-time premium) into a pool account. The payment to the pool is meant to cover (together with similar payments from other bank operators) the aggregate risk of failures calculated for all participants. If a bank fails and the operator defaults, the site-specific bond is used first for the rehabilitation work and the bond pool pays for all the excess not covered. The advantage of a pool is that it can reduce costs to individual operators while still providing an incentive to perform. The difficulty is in setting an appropriate fee or premium given uncertainties in predicting the likely number of bank failures and the costs of correcting them.

Some regulatory schemes allow self-bonding or third-party corporate guarantees. Corporate guarantees can be as effective as sureties if the solvency of the guarantor can be continuously monitored *and* provided that the regulatory authority can quickly access the guarantee funds without substantial litigation. They are poor substitutes if these factors are not present. Self-bonding, sometimes called the "financial test," is a less reliable guarantee. Bonds, trust funds, and other formal financial instruments are meant to protect the public interest in the event of a default by the credit producer; in contrast, the financial test essentially assumes that no default will occur based on the size or assets of the credit producer. If this assumption is wrong, or the company denies responsibility for a failure, the regulator is no better off than if it had no such guarantee; it will need to file suit, or may need to attempt to extract assets from a bankruptcy without an enforceable security interest that would give it a priority claim.

Government-operated banks often maintain that they should not be subject to financial assurance requirements. Although, presumably, government agencies exist in perpetuity and have the financial credit of the state, local, or federal government behind them, in reality financial difficulties are endemic to governmental agencies. Appropriations may not be made by the legislature to meet obligations that are perceived as nonessential; or, funding priorities may shift. In short, absent a designated source of committed funds, government-operated banks may be even less reliable than some private banks. For example, the Louisiana Department of Transportation (LDOT) bank has suffered from the absence of a trust fund or similar instrument. Eighty-three percent of the credits in the bank were to have been generated by management of the land to enhance wildlife habitat. Although most of the credits were used, the management activities did not occur. The site owner—the State Department of Wildlife and Fisheries—received no funding from its own appropriations or from the DOT to conduct these management activities.

If a formal financial instrument is used, it may be funded in several ways. The banking program may simply require a financial instrument to be posted in a given amount, leaving it to the credit producer to recoup this expense in the marketplace. Alternatively, the banking plan can assess a fixed surcharge on each sale of credits. This approach guarantees the fund a certain amount and also links the increase in the fund to the size of the risk at issue; as more credits are issued and relied on, more funding is available to handle failures.

How long should a bonding requirement, trust fund, and the like be in effect? This question has no fixed answer. For example, because banked mitigation wetlands are designed to compensate for a wetland that conceivably would have existed for decades—if not in perpetuity—it is not unreasonable from an ecological perspective to require a perpetual care fund. On the other hand, many types of wetlands do not require long-term care; indeed, if they are truly replacing functions and values, they should, by definition, be self-sustaining. Thus, financial assurance could be for a far more limited duration.

The most logical approach is to link financial assurance requirements to the "ecological success" criteria established for the bank. The guarantee for a successful bottomland hardwood creation project, for example, would need to be for a far greater length of time than that for a restored prairie pothole. Linking assurance to success criteria also provides an incentive for self-monitoring and speedy correction of problems by the credit producer or other responsible entity. The length of time for financial assurances to remain in effect could be determined either (1) by linking it to the achievement of site-specific success criteria, or (2) by establishing a fixed period by wetland and mitigation type, with a provision for release only upon demonstration of success at the end of that period.

The strongest guarantor of success, of course, and one that obviates the need for after-the-fact enforcement, is advance mitigation. The Aciquia Mitigation Bank in Idaho, for example, requires that mitigation must be complete and successful before credits can be earned and does not allow debits to be made until the credits are earned. Other banking schemes attempt to use interim milestones as a partial substitute for complete advance mitigation. The Weisenfeld bank permit makes the compensation ratio for credits dependent on the successive achievement of six specific success criteria. The LDOT bank is an example of what can go wrong when a developer is given credits in advance of promised work. The LDOT received 64 annual available habitat units for purchasing (preserving) 3,000 acres of wetlands. It needed more credits so it agreed to actively manage and enhance the wildlife habitat of the land in exchange for receiving an additional 300 credits. The LDOT has currently debited all but 70 credits without undertaking any management activities. There has been no habitat enhancement and there is no formal banking agreement through which to enforce the obligation.

Some banks have anticipated this type of problem and have provisions in their banking instruments for the revision of credits after reviewing monitoring reports. The MOA for Prince George's County Bank in Maryland, for example, authorizes the oversight team to recommend revising the credits and debits of the bank, as well as the MOA, after reviewing their monitoring reports. Indeed, it is conceivable that a bank could be penalized not just monetarily for a violation, but through forfeiture of certain credits, or even bank lands, to the state or federal regulatory authority.

Forfeiture of financial assurance funds can also provide a powerful enforcement tool. The bank agreement for the ACWHEP bank in California, for example, required the Mission Viejo Company to provide an $800,000 bond to Orange County at the start of the construction of Phase I to ensure the success of habitat value replacement. The proposed Homebuilders Association of Greater Chicago Mitigation Bank MOA calls for money to be kept in an escrow account. In Georgia, the Millhaven bank must maintain a bond to guarantee monitoring and maintenance of the bank for a five-year period. The key to using a financial instrument as an enforcement tool is that the regulatory agency must be able to access it without a prolonged legal process. The agency should be able to draw on the fund, bond, or letter of credit upon its determination that a violation has occurred and not been remedied. If the agency's ability to access the funds is contingent upon winning a court case, the enforcement utility of the instrument is significantly reduced.

Prospects for Banking

Wetland mitigation banking offers a significant opportunity to make compensatory mitigation more effective ecologically while offering economies and certainty to developers. To the extent that mitigation wetlands are banked in advance of wetland conversions, banking can also provide timing advantages over traditional on-site mitigation. Nevertheless, wetland mitigation banking also offers some uncertainties as well. Because the market for credits is driven by regulatory-created demand, it is uncertain whether mitigation banks will succeed in attracting private capital. Entrepreneurial commercial banks have been slow to get underway. To the extent to which wetland regulation itself remains under attack, banking will remain an uncertain enterprise. Given this uncertainty, the natural tendency of those seeking to start entrepreneurial banks will be to seek the assistance of government in providing them with a steady market—thus leading to calls for the elimination of "sequencing," also disliked by many developers. Sequencing now requires the avoidance and minimization of impacts on-site before allowing compensatory mitigation.

However, sequencing can serve valuable ecological functions in many settings by preserving the operation of functioning wetland systems in specific places on the landscape. A wholesale reliance on wetland mitigation banks in lieu of other wetland protection tools would be as inappropriate, in its own way, as a blanket prohibition on wetland banking. The most ecologically sensitive approach would recognize a combination of sequencing, on-site mitigation, and off-site mitigation banking.

Wetland mitigation banking is likely to be an effective tool for wetland conservation only if regulators give equal attention to on-site compensatory mitigation. If regulators promulgate elaborate permit requirements for wetland banks, with bonding, monitoring, enforcement, and other provisions, while doing nothing for on-site mitigation, wetland mitigation banking will be disadvantaged while on-site mitigation will freely continue its checkered career. On-site mitigation and banking need not be subject to identical standards, but they do need parity of treatment.

Notes

1. Portions of this chapter have been adapted from *Wetland Mitigation Banking*. Copyright 1993, Environmental Law Institute, all rights reserved. Used by permission.
2. Institute for Water Resources, *Briefing Book: Mitigation Banking Concepts*. Washington, DC, October 1983.
3. Cathleen Short, "Mitigation Banking," *Biological Report* 88(41). U.S. Department of the Interior, July 1988.
4. Environmental Law Institute, *Wetland Mitigation Banking*. Washington, DC, 1993.
5. Kevin L. Erwin, *An Evaluation of Wetland Mitigation Within the South Florida Water Management District*. Fort Myers, FL: South Florida Water Management District, 1991.
6. National Research Council, *Restoration of Aquatic Ecosystems: Science, Technology and Public Policy,* Washington, DC: National Academy of Science, 1992.
7. "Federal Guidance for the Establishment, Use and Operation of Mitigation Banks," *Federal Register* 60(43): 12292, March 6, 1995.

3

Federal Wetland Mitigation Policies

John Studt and Robert D. Sokolove*

Federal regulation of discharges into the nation's waterways began in 1899 with the enactment of the Rivers and Harbors Act of 1899.[1] Section 10 of the act prohibits dredging or discharging material in the navigable waters of the United States without a permit from the U.S. Army Corps of Engineers. It was not until the late 1960s and early 1970s, however, that the Corps's jurisdiction was expanded beyond simply maintaining the navigability of the nation's waters for interstate commerce to the regulation of all U.S. waters, including wetlands.

In 1968, under pressure from the public and the U.S. Fish and Wildlife Service (FWS), the Corps revised its permitting process under the Rivers and Harbors Act to include the consideration of environmental values. This became known as the "public interest review." The purpose of this review is to balance the reasonably foreseeable adverse impacts of a project, such as habitat destruction and pollution, with its positive impacts, such as economic development, jobs, and tax revenues. In 1969, the National Environmental Policy Act (NEPA) was enacted,[2] which further expanded the scope of the Corps's public interest review.

*Lindell L. Marsh contributed to this chapter.

It was not until 1972, however, with the enactment of the Federal Water Pollution Control Act, more commonly known as the Clean Water Act,[3] that the Corps's jurisdiction was expanded to include wetlands. Numerous federal and state laws can restrict development in wetlands, but Section 404 of the Clean Water Act has the greatest impact on such development.[4]

The Clean Water Act was enacted to restore and maintain the chemical, physical, and biological integrity of the nation's waters. Under Section 404(a) of the act, a permit is required from the Corps before dredged or fill material can be discharged into the waters of the United States, including wetlands.[5] However, the act provided no guidelines to evaluate such permit applications. Therefore, the Corps relies on its public interest review and guidelines developed by EPA under Section 404(b)(1) of the act.[6] These guidelines, known as the (b)(1) guidelines, state that "no discharge of dredged or fill material shall be permitted if there is a practicable alternative to the proposed discharge which would have less adverse impact on the aquatic ecosystem."[7] Moreover, if the proposed development is not "water dependent," that is, does not require access or proximity to an aquatic site to fulfill its basic purpose, then practicable alternatives are presumed to exist, unless clearly demonstrated otherwise. In addition, under Section 404(c), EPA may veto a permit issued by the Corps if a proposed project will have "unacceptable adverse effect[s] on municipal water supplies, shellfish beds and fishery areas (including spawning and breeding areas), wildlife, or recreational areas."[8] EPA vetoes are rare, however. Since the program started, EPA has issued less than a dozen vetoes out of an estimated 160,000 permit applications received by the Corps.

When the Clean Water Act was enacted, the definition of "waters of the United States" was left up to the Corps and EPA. Initially, these agencies interpreted the term narrowly to encompass only traditional navigable waters. This definition was broadened in 1975, following a successful legal challenge by the Natural Resources Defense Council, to include navigable waters and their tributaries, interstate waters and their tributaries, nonnavigable intrastate waters whose use or misuse could affect interstate commerce, and all freshwater wetlands adjacent or connected to other waters protected by the statute.[9] Today, the Corps defines wetlands as "those areas that are inundated or saturated by surface or groundwater at a frequency and duration sufficient to support, and that under normal circumstances do support, a prevalence of vegetation typically adapted for life in saturated soil conditions."[10] Under this definition, wetlands include marshes, bogs, swamps, and similar areas. However, they may also include areas that would not typically be considered wetlands.

In 1993, the Corps jurisdiction was broadened further when it promulgated what is known as the Excavation Rule. The rule expanded the definition of "discharge" and closed a large number of loopholes that allowed the lawful destruction of wetlands every year.[11] This rulemaking was initiated in response to a settlement agreement reached between the National Wildlife Federation and the Corps after the Corps refused to claim jurisdiction over a development project that was clearly evading the 404 permit requirements by removing the water and vegetation from a wetland to render a site dry.[12] Prior to the Excavation Rule, many activities that destroyed wetlands were outside the scope of the Clean Water Act: one could legally drain, dig ditches through, or dig large holes in a wetland without a permit so long as none of the dirt, mud, or sand was deposited in the wetland. The Excavation Rule changed the focus of this evaluation to the environmental impact of all such activities in or around wetlands to determine if they destroy or degrade the wetlands. Therefore, even if a person never operates within the confines of a wetland, a Section 404 permit may still be required if the activities destroy or degrade a wetland. The Excavation Rule does not regulate the incidental addition of dredged material that does not destroy or degrade a wetland. However, the burden of showing this is on the landowner.[13]

Whether a parcel of land is classified as a wetland depends upon how it is delineated.[14] Prior to 1989, the Corps, EPA, U.S. Fish and Wildlife Service (FWS), and Soil Conservation Service each had its own method for delineating wetlands. Not surprisingly, this often resulted in four different delineations for the same site. In 1989, these four agencies jointly developed a "Federal Manual for Identifying and Delineating Jurisdictional Wetlands." This manual, however, came under heavy attack for expanding the Corps's jurisdiction over areas not previously considered wetlands. In 1991, the White House proposed revisions to the 1989 manual that would have substantially reduced the scope of the Corps's jurisdiction. These revisions also came under heavy attack. In 1992, President Bush signed an appropriations rider to the Energy and Water Resources Development Act that prohibited the Corps from expending any funds to perform wetland delineations using the 1989 manual. Today, the Corps is using the wetland delineation manual it developed in 1987. The EPA also adopted the Corps's 1987 manual in 1993. This manual will be utilized until the National Academy of Sciences, in response to a Congressional mandate, finalizes a scientific analysis on wetland delineation methods. This debate has resulted in a high level of uncertainty as to whether the Corps's jurisdiction will extend over any particular parcel of property, and it ensures that the question of whether a development or other project will require mitigation will not be resolved with any certainty in the near future.

Not all wetland impacts, however, require a Section 404 permit. For example, Section 404(e) authorizes the Corps to issue general permits on a state, regional, or nationwide basis for certain categories of activities in wetlands that are "similar in nature, [and] will cause only minimal adverse effect to the environment."[15] General permits are a kind of generic permit that grants blanket authorization for certain types of fill. General permits issued nationwide are called, appropriately, nationwide permits.

If an activity is covered by the Corps's general permit program, no Section 404 permit application need be filed unless the state in which the impact will occur has not issued a water quality certification or coastal program consistency determination for the general permit.[16] There are also a number of statutory exemptions under Section 404(f)(1).[17] However, these exemptions are not applicable if the purpose of the activity is to "recapture" an area of navigable waters into a new use or where the flow or reach of the waters have been impaired or reduced.[18]

Mitigation Under Other Federal Laws

Since 1972, the Clean Water Act has evolved into the major federal program regulating activities in the nation's wetlands. Other federal acts also affect activities in wetlands and endangered species habitat and require mitigation to offset adverse environmental impacts. The relevant federal laws are outlined briefly herein, followed by a discussion of the mitigation requirements and policies under the Clean Water Act.

The Fish and Wildlife Coordination Act

One of the earliest federal statutes to require mitigation for habitat loss is the Fish and Wildlife Coordination Act,[19] enacted in 1934 and strengthened by amendments in 1946, 1958, and 1965. The act applies to both congressionally authorized and federally permitted "water resource development projects," and specifically to issuance of Section 404 permits. It requires the Corps to "consult" with FWS and to consider FWS's recommendations for avoiding or compensating for habitat loss, but it does not require the Corps to adopt those recommendations. FWS has developed and published its own comprehensive mitigation policy.[20] The FWS policy creates four resource categories and ranks habitat according to its scarcity value, with "unique and irreplaceable" habitat receiving highest priority. The policy then prescribes a mitigation planning goal ranging from "no loss of existing habitat value" to "minimize loss of habitat value."

The National Environmental Policy Act (NEPA)

Enactment of the National Environmental Policy Act[21] in 1969 ushered in a new era of environmental planning and ecological stewardship. NEPA establishes a judicially enforceable procedural requirement for federal agencies to prepare an environmental impact statement to consider the environmental effects of "major federal actions," including actions like issuing Section 404 permits. Among the factors to be considered are alternatives to the proposed action and ways to minimize any resulting harm. In 1978, the Council on Environmental Quality (CEQ) published regulations, binding on all federal agencies, that spelled out the procedures required to implement NEPA, including mitigation responsibilities. The CEQ regulations defined "mitigation" as

1. avoiding the impact altogether by not taking a certain action or parts of an action,
2. minimizing impacts by limiting the degree or magnitude of the action and its implementation,
3. rectifying the impact by repairing, rehabilitating, or restoring the affected environment,
4. reducing or eliminating the impact over time by preservation and maintenance operations during the life of the action,
5. compensating for the impact by replacing or providing substitute resources or environments.

Like the Fish and Wildlife Coordination Act, NEPA is a procedural statute; it does not require that agencies adopt any mitigation measures at all. However, the process motivates agencies to incorporate mitigation conditions into permit decisions.

Endangered Species Act of 1973 (ESA)

The Endangered Species Act sets forth a strong national mandate to protect and manage endangered species and their habitats, rare animals, and plants that are in danger of extinction. The act prohibits any person from "taking" an endangered species. Taking includes hunting, trapping, harming, or harassing such species.

Historically mitigation banking has focused on mitigation required under the Clean Water Act. Adverse impacts to endangered species habitat were permitted, in effect, through consultation by federal agencies with the FWS in cases where a federal action was involved. There was no procedure for the direct permitting of development in endangered species habitat by the private sector that were unrelated to a federal action (e.g., such as a Section 404 permit). In addition there was a relatively unused provision under Section 4(d) of ESA, whereby the Secretary of Interior could promulgate a "special rule" that would prescribe the circumstances when a "threatened" species could be taken or not taken.

In 1982, Congress amended the ESA to authorize the FWS to issue a permit for "incidental taking" of endangered species habitat, providing that a permit applicant prepare a "habitat conservation plan" (HCP) that, among other things, "avoided and mitigated the take of the species to the fullest extent practicable" and did not "appreciably reduce the likelihood of the survival and recovery of the species." The result has been a growing emphasis nationwide on the preparation of HCPs that balance the conservation of listed species with limited development in endangered species habitat. Further, in recognition of the fact that the permitting provisions of ESA are invoked when a species is on the brink of extinction, the HCPs prepared have increasingly been focused on a broad spectrum of species and habitats. As a result, there has been a growing use of mitigation banks to provide for the compensation for wildlife impacts in upland areas as well as in areas covered by the Clean Water Act.

Food Security Act (FSA) of 1985, Amended in 1990[22]

Known commonly as the Farm Bill, the act removes some of the incentives to convert wetlands to farmland. The act does not prohibit conversions; farmers can still convert wetlands after obtaining a Section 404 permit from the Corps, but they will be ineligible, with some exceptions, for U.S. Department of Agriculture (USDA) benefits, such as crop price supports, disaster payments, and Farmers Home Administration loans.

Farmers violating the so-called "Swampbuster" provisions of the act can lose USDA benefits until the converted wetland is restored.

Wetland Mitigation Requirements under the Clean Water Act

Dating back at least to the mid-1930s the Corps's concept of mitigation included avoiding impacts to the environment. For example, in 1933 the Corps denied a permit for a wharf in the Potomac River near Washington, D.C., because alternatives existed that would be less damaging to the overall aesthetic environment of the Virginia shoreline of the river.

Among the earliest applications of the mitigation concept were efforts to compensate for the effects of dams on anadromous fish populations through the construction of fish passages or fish hatcheries. Throughout the 1970s, the compensatory mitigation concept was expanded from protecting a single species to protecting entire habitats, primarily aquatic or wetland ecosystems. Many of these early mitigation programs were performed in response to federal activities undertaken pursuant to the Rivers and Harbors Act and the Clean Water Act.[23]

Prior to 1989, the Corps took the position that compensatory mitigation could be used to satisfy the public interest test and other legal requirements for the issuance of a permit. This approach resulted in decisions that sometimes included compensatory mitigation for impacts even if the impacts could have been avoided. In many cases, acceptance of compensatory mitigation where impacts were avoidable involved the creation of new wetlands. Unfortunately, many of these created wetlands fared poorly, due to inadequate design, poor construction techniques, and a lack of monitoring and maintenance.

The amount of compensatory mitigation required depended on the impacts to the aquatic environment, when balanced with the public benefits of the proposed project, and whether the proposed mitigation was "practicable" to an applicant. In addition, the Corps factored into its permit decisions the belief that, generally, mitigation should be close by (on-site), or at least within the same watershed, and typically should provide similar aquatic resource values (in-kind) as the wetland being "lost" to development.

EPA, in contrast, argued that compensatory mitigation should be the last option in a sequence of possible alternatives that should be examined after the Corps's public interest review and the EPA's Section 404(b)(1) guidelines have been satisfied.[24] Under the EPA's "sequencing" mitigation guidelines, off-site compensatory mitigation for unavoidable wetland impacts is acceptable only after it is determined that (1) the potential adverse impacts to the wetlands have been avoided to the maximum extent practicable, (2) appropriate and practicable measures have been taken to minimize the remaining unavoidable adverse impacts, and (3) appropriate and practicable actions have been taken to compensate for unavoidable adverse impacts. In other words, avoid first, restore or enhance second, and after other options are exhausted, create new wetlands to replace those lost to development. The EPA's sequencing guidelines were adopted by the Corps in a Memorandum of Agreement (MOA) between the Corps and EPA dated November 15, 1989, but effective February 7, 1990 ("Mitigation MOA").[25] The MOA codified the mitigation sequencing requirement. Thus, permit applicants must demonstrate that they have made every reasonable effort to avoid and minimize wetland losses through careful location and design before compensatory mitigation techniques such as wetland restoration, creation, or enhancement will be considered. Compensatory measures must be "appropriate and practicable." "Appropriate" mitigation is based on the ecological value of the affected wetland. "Practicable" is defined in Section 230.3(q) of the guidelines and requires consideration of "cost, existing technology, and logistics in light of overall project purposes." The Mitigation MOA requires a Section 404 permit applicant to replace the functional value of the wetlands being impacted at a ratio consistent with the policy of no net loss and with an adequate margin of safety to reflect the expected degree of success associated with the permittee's mitigation plan.[26] A replacement ratio of less than 1:1 (one acre created for every acre impacted) is permissible where the functional values of the impacted wetlands are low and the likelihood of the proposed mitigation succeeding is high.[27] Moreover, mitigation is not required where it is determined that:

> the mitigation measures necessary to meet this goal [of no net loss] are not feasible, not practicable, or would accomplish only inconsequential reductions in impacts. Consequently, it is recognized that no net loss of wetlands functions and values may not be achieved in each and every permit action.

The MOA also provides that mitigation is not necessary if the proposed discharge is necessary to avoid environmental harm or can reasonably be expected to result in environmental gain or insignificant impact.[28]

In order to achieve the goal of no net loss of wetlands, the Mitigation MOA established a preference for "in-kind" mitigation (i.e., replacing impacted wetlands with wetlands that have the same functional values) over "out-of-kind" mitigation. Functional values of impacted wetlands are determined utilizing "aquatic site assessment techniques generally recognized by experts in the field and/or the best professional judgment of federal and state agency representatives."[29] The two aquatic assessment techniques most widely used are the Habitat Evaluation Procedure (HEP) and the Wetland Evaluation Technique (WET).[30] The MOA also states a preference for "on-site" versus "off-site" mitigation using restoration techniques over creation.[31] Only after a Section 404 permit applicant has complied with the just mentioned sequencing rules by showing that on-site mitigation is not practicable will the permittee be allowed to mitigate for its unavoidable wetland impacts through the use of off-site mitigation. If "off-site" mitigation is permitted, the Mitigation MOA provides that it should be undertaken "in close proximity and to the extent possible in the same watershed."[32] The MOA also accepted the concept of wetland mitigation banking, provided that use of the bank meets the objectives of the MOA.

The Mitigation MOA governs individual permits, including after-the-fact permits, for which applications were filed after February 7, 1990. It does not apply to general permits, including nationwide permits. It specifies a clear preference for on-site, in-kind replacement of wetland functions and values and establishes a minimum one-to-one ratio as a rule of thumb for replacement. Mitigation banks are recognized as an "acceptable form of compensatory mitigation under specific criteria designed to ensure an environmentally successful bank." However, the MOA notes that simple purchase or "preservation" of existing wetlands will not be considered adequate compensation except in "exceptional circumstances."

The mitigation MOA also provides for deviations from the sequencing requirements when the requirements have been incorporated into a comprehensive plan approved by the Corps and EPA.

The Corps's regulatory decisions generally will follow a plan, such as a watershed management plan, that should identify wetland areas within the watershed that could be restored, enhanced, created, or preserved as mitigation for wetlands impacts in the same watershed. Such priority mitigation areas could be incorporated into mitigation banks.

Through its Programmatic General Permits (PGP), the Corps may encourage the development of watershed management plans. A PGP is a type of general permit that is developed by the Corps based on a strong state, local, or regional program that protects the aquatic environment. The PGP provides for a substantial reduction in duplication between the Corps regulatory program and the state or local regulatory program. A PGP also provides the state, local, or regional regulatory authority with greater flexibility in defining the specific role the state or local organization wants to establish. At the same time, a PGP provides many environmental safeguards to ensure the environment is protected to the level provided by the Corps regulatory program.

Under a PGP, the Corps relies on state or local agencies to review permit applications, thus reducing duplication among agencies and expediting the permit process. When state or local agencies approve a project, the Corps provides its approval quickly, unless there is some element of the federal interest that requires additional review and attention.

Programmatic General Permits provide several features to ensure that the aquatic environment is protected to the level provided by the Corps regulatory program. A Corps PGP would contain specific conditions that protect the aquatic environment and the Corps can add special conditions to any specific authorization that it issues under the PGP. Moreover, under its discretionary authority, the Corps can always require any particular activity that could be authorized by the PGP to be reviewed under the Corps's individual permit evaluation process.

The PGP is issued in the same manner as an individual permit. That is, the proposed PGP is published for public comment through a Corps public notice and all comments received are considered before the Corps decides to issue the PGP. The proposed PGP would consist of a description of the program that the Corps is proposing to cooperate with, the terms under which a Corps authorization would be issued, and special conditions that an authorized project must meet. The terms of the PGP would specify the geographic location covered by the PGP, any activities or geographic areas excluded from coverage, and the manner in which authorizations would be conveyed (e.g., whether a Public Comment Notice would be required). The special conditions would establish certain requirements for authorization, including compensatory mitigation requirements. If a mitigation bank is established, conditions of the PGP could direct use of the mitigation bank to provide compensatory mitigation for some or all activities authorized under the permit.

Once a PGP is issued, then activities that the Corps determines meet the terms and conditions of the PGP are authorized in an expedited manner. The specific authorizations issued by the Corps under the PGP often would include additional project-specific special conditions to protect the environment.

In addition to PGPs, other mechanisms aimed at guiding or coordinating land use and land management decisions include Advance Identification (ADID) under Section 404 of the Clean Water Act and Special Area Management Plans (SAMPs) under the Coastal Zone Management Act. Besides providing greater predictability and certainty to the regulatory process, these planning mechanisms can help educate participants to the importance of protecting natural resources such as wetlands. In addition, such planning mechanisms can help ensure consistency among local, state, and federal land use and planning decisions. Chapter 8 provides a more detailed discussion of mitigation banks and watershed planning.

The concepts of in-kind and on-site mitigation have changed over the past several years. Generally, the Corps believes that replacement aquatic resources should be of the same general type and in the same general area as the resources adversely altered by development. In this way, the compensatory mitigation can be viewed as "replacing" the aquatic values lost through project development. This general philosophy remains a part of the Corps's regulatory program. In many cases, however, wetland or other aquatic resources lost in an area are not the type most needed within the watershed. The Corps general approach is to require compensatory wetland mitigation of the type and location that is needed most in the watershed and located where it will do the most good for the watershed. If a watershed management plan exists, this determination is easier.

Wetland Mitigation Banking Guidance Documents

Several EPA, Corps, and FWS regional offices have drafted or issued guidelines for establishment and operation of wetland mitigation banks as an acceptable form of compensatory mitigation.

EPA Region IX issued final guidelines on December 20, 1991. These guidelines referenced the Mitigation MOA and reinforced the requirement that all impacts must be avoided or minimized before compensatory mitigation is considered. They also identified specific situations where compensatory mitigation in the form of mitigation banking is appropriate. These include water-dependent projects; projects involving small unavoidable impacts; linear projects such as highways, which involve many minor impacts; and routine repair and maintenance of public structures, such as the cleaning of drainage ditches.

EPA Region IV released draft guidelines in 1992 similar to those issued in final form by Region IX. The most significant difference is that the Region IV guidelines include activities authorized under general permits in the list of projects generally appropriate for mitigation banking. EPA Regions I, II, and III developed draft guidelines jointly with the Corps's North Atlantic Division and New England Division, Region V of the FWS, and the Northeast Region of the National Marine Fisheries Service. The purpose of this document is to provide guidance on the development and operation of wetland mitigation banks associated with highway construction. Once again, the agencies cited the 1990 Mitigation MOA and required that sequencing be applied before mitigation banking is considered. Neither of these draft guidelines was finalized.

In 1993, the Galveston and Omaha district offices of the Corps of Engineers issued final guidelines for the use of wetland mitigation banks. The Omaha guidelines expressly provide that all projects must follow the five-step sequencing requirement set forth in the CEQ regulations and incorporate sequencing from the Section 404(b)(1) guidelines. The Galveston guidelines adopt the "avoidance, minimization, and compensation" language of the Mitigation MOA.

On August 23, 1993, the Corps and EPA issued a Regulatory Guidance Letter ("Mitigation Banking RGL"),[33] in conjunction with a White House policy statement on protecting America's wetlands.[34] Both the Mitigation Banking RGL and the White House Wetlands Policy Statement endorsed the concept of private, entrepreneurial wetland mitigation banking as a form of off-site compensatory mitigation. Prior to this, mitigation banking was viewed as a creature of the public domain. The Mitigation Banking RGL restates the agencies' preference for on-site, in-kind compensatory mitigation. However, it also provides detailed guidance on how to establish private, entrepreneurial wetland mitigation banks. The RGL defines wetland mitigation banking as:

> the restoration, creation, enhancement, and, in exceptional circumstances, preservation of wetlands or other aquatic habitats expressly for the purpose of providing compensatory mitigation in advance of discharges into wetlands permitted under the Section 404 regulatory program.[35]

The RGL still requires a permittee to go through the complete Section 404 permitting process.[36] However, once it is determined that off-site mitigation is warranted, the RGL allows a Section 404 permit applicant to use a wetland mitigation bank, preferably one within the same watershed as the impact site, to compensate for the wetland losses.[37] Whether a particular mitigation bank is appropriate for mitigating for an unavoidable wetland loss depends upon the wetland functions, landscape position, and affected species populations at both the impact and bank sites.[38]

Federal Mitigation Banking Guidance

On November 28, 1995, the Corps, EPA, Natural Resources Conservation Service (NRCS, formerly the Soil Conservation Service), FWS, and NMFS issued guidance (Guidance) regarding the establishment, use, and operation of mitigation banks.[39] In the Guidance, the agencies recognized the potential benefits that mitigation banking offers for streamlining the permit process and providing more effective mitigation and encouraged the establishment of such banks. The Guidance uses virtually the same definition of mitigation banking as the RGL.

Citing the Section 404 (b)(1) and the Council of Environmental Quality Guidelines, the Guidance confirms that "the use of credits may only be authorized for purposes of complying with Section 10/404 when adverse impacts are unavoidable" and that "credits may only be authorized when on-site compensation is either not practicable or use of a mitigation bank is environmentally preferable to on-site compensation." The last phrase suggests that it may be possible to establish the preference for off-site mitigation in appropriate circumstances.

The Guidance states that typically the Corps will serve as the lead agency for the establishment of mitigation banks except where the bank is solely for the purpose of complying with the Swampbuster provisions of Food Security Act (FSA), in which case NRCS would be the lead agency.

Mitigation Banking Instruments

The Guidance states that "all mitigation banks need to have a banking instrument as documentation of agency concurrence on the objectives and administration of the bank." The banking instrument should describe in detail the "physical and legal characteristics of the bank, and how the bank will be established and operated." Bank sponsors will be responsible for assuring the success of the wetlands that were created, restored, enhanced, or preserved when the bank was established and that "it is extremely important that an enforceable mechanism be adopted" to hold bank sponsors accountable.[40]

Significantly, the Guidance does not characterize the "banking instrument" as an "agreement" binding upon all of the parties, although it suggests that it "will be signed by the bank sponsor and resource agencies represented on the Mitigation Bank Review Team," and that "in signing a banking instrument, an agency agrees to the terms of that instrument."[41]

The Guidance states that the following information should be addressed, "as appropriate" within the banking instrument:

1. Bank goals and objectives
2. Ownership of bank lands
3. Bank size and classes of wetlands and/or other aquatic resources proposed for inclusion in the bank
4. Description of baseline conditions
5. Geographic service area
6. Wetland classes or other aquatic resource impacts suitable for compensation
7. Methods for determining credits and debits
8. Accounting procedures
9. Performance standards
10. Reporting protocols and monitoring plan

11. Contingency and remedial actions and responsibilities
12. Financial assurances
13. Compensation ratios
14. Provision for long-term management and maintenance[42]

In cases where the establishment of the mitigation bank involves a discharge requiring Section 10/404 authorization, the banking instrument would be included in the Corps's permit. Where the bank is established pursuant to FSA, the banking instrument would be included in the plan developed or approved by NRCS and FWS.

The Guidance contemplates the establishment of a Mitigation Bank Review Team chaired by the Corps (except where only the provisions of FSA are being addressed, in which case the NRCS would chair the review team). Members of the review team would sign the banking instrument, although signature would not be required. Normally these agencies would include the Corps, EPA, FWS, NMFS, and NRCS but could also include states, tribes, local agencies, or other agencies where the participant has an authority or mandate directly affecting or affected by the establishment, use, or operation of a bank.

The primary role of the review team will be to facilitate the establishment of mitigation banks through the development of mitigation banking instruments. The review team "will strive to obtain consensus on its actions."[43]

The Corps and NRCS, respectively, are given the authority for authorizing the use of a bank in a particular situation and determining the number and availability of credits in accordance with the terms of the banking instrument.[44]

The Guidance contemplates collaborative processing and decisions by consensus. However, if consensus cannot be reached within a reasonable time or if an agency member of a review team considers that a particular decision raises concern regarding the application of existing policy or procedures, an agency may request review by a higher level within each agency. If resolution is still not achieved any agency may initiate interagency review through written notification to the other agencies. Within 20 days the district engineer or state conservationist, as appropriate, will "lead necessary discussions to achieve interagency concurrence on the issue of concern." The bank sponsor may also request review by the district engineer or state conservationist, as appropriate, if it believes that inadequate progress has been made.

During the permit process incorporating a banking instrument, any agency on a review team can initiate the review of issues regarding whether a proposed use may not comply with the terms of a banking instrument. Following the issuance of a permit or permits, an agency can raise questions through the review team chair as to whether an issued permit (or permits) "reflects a pattern of concern regarding the application of the terms of the banking instrument."

Each of the guidelines and guidance documents identifies important issues for consideration in the structuring of wetland mitigation banks.

The use of wetland mitigation banks is not without precedent. For years, developers have utilized, and the government has embraced, the concept of transferable development rights (TDRs).[45] Under the TDR concept, developers are permitted to increase building densities in more urban, developed portions of communities in exchange for easements setting aside greater acreage of open space in more rural or agricultural areas. This balancing has been viewed by some as an important vehicle to provide increased growth in areas more economically advantageous to a developer, while providing necessary open space for the well-being of the community. As with wetland mitigation banking, however, some critics charge that TDRs are little more than "checkbook zoning," allowing a developer to increase building density to the detriment of some and, presumably, to the benefit of others.

Notes

1. 33 U.S.C. §§ 401 et seq.
2. 42 U.S.C. §§ 4321 et seq.
3. 33 U.S.C. §§ 1251 et seq.
4. 33 U.S.C. § 1344.
5. 33 U.S.C. § 1344(a).
6. 33 U.S.C. § 1344(b)(1); 40 CFR Part 230.
7. 40 CFR 230.10(a).
8. CWA Section 404(c), 33 U.S.C. Section 1344(c).
9. 33 CFR 328.3(a).
10. 33 CFR 328.3(b); 40 CFR 230.3(t). This regulatory definition was upheld by the Supreme Court in *United States v. Riverside Bayview Homes,* 474 U.S. 121 (1985).
11. 33 CFR 323.2.
12. *North Carolina Wildlife Federation v. Tulloch,* Civil No. C90-713-CIV-5-30 (E.D.N.C. 1992).
13. 3 CFR 323.2(d)(3)(i).
14. Wetland delineation is the process by which the existence and location of wetlands on a property are determined.

15. 33 U.S.C. § 1344(e)(1); 33 CFR 320.1.
16. 33 CFR 330.1.
17. 33 U.S.C. § 1344(f)(1); 40 CFR 232.3.
18. 33 U.S.C. § 1344(f)(2).
19. 16 U.S.C. § 661-667e.
20. 46 Fed. Reg. 7644 (Jan. 23, 1986).
21. 42 U.S.C. §§ 4321-4335.
22. The Food Security Act of 1985, (P.L. 99-198, 99 Stat. 1354) as amended by the Food, Agriculture, Conservation and Trade Act of 1990.
23. Want, William L., *Law of Wetland Regulation* § 6.10[2][a] (Clark Boardman 1994).
24. Id.
25. Memorandum of Agreement between the Environmental Protection Agency and the Department of the Army Concerning the Determination of Mitigation under the Clean Water Act Section 404(b)(1) Guidelines (Nov. 15, 1989) ("Mitigation MOA").
26. Mitigation MOA at Part III.B.
27. Mitigation MOA at Part III.B.
28. Mitigation MOA at Part II.C.
29. Mitigation MOA at Part III.B.
30. The HEP is a species-specific approach to impact and resource assessment based on the assumption that it is possible to numerically describe habitat quality and quantity. This numerical habitat analysis translates into a mitigation ratio describing the wetland area that must be replaced for each acre of wetland impacted. The WET is similar in nature, but incorporates many factors not addressed by the HEP that are wetland-specific.
31. Mitigation MOA at Part II.C.3.
32. Id.
33. Regulatory Guidance Letter, No. 93-2, Establishment and Use of Wetland Mitigation Banks In The Clean Water Act Section 404 Regulatory Program (1993) ("Mitigation Banking RGL").
34. Protecting America's Wetlands: A Fair, Flexible, and Effective Approach (White House Office of Environmental Policy 1993) ("White House Wetlands Policy Statement").
35. Mitigation Banking RGL at ¶ 2.
36. Mitigation Banking RGL at ¶ 3.
37. Mitigation Banking RGL at ¶ 4.

38. Id.
39. 60 FR 58605-58614, November 28, 1995.
40. Federal Proposed Guidance, p. 58608.
41. Ibid, p. 58610.
42. Ibid.
43. Ibid.
44. Id, p. 58610.
45. The sale and purchase of wetland mitigation credits is also analogous to the sale and purchase of air credits under the Clean Air Act, 42 U.S.C. §§ 7401, et seq.

4

State Mitigation Banking Programs: The Florida Experience

Ann Redmond, Terrie Bates, Frank Bernadino, and Robert M. Rhodes

Most states have enacted laws that apply specifically to wetlands. The laws vary from those that authorize states to acquire and preserve wetlands to those that require permits for construction in wetlands. For example, Oregon requires permits for both filling and removal activities in all waters of the state including freshwater and tidal wetlands. Michigan and New Jersey, states that have assumed implementation of the federal permit program under the Clean Water Act Section 404(g), require permits for a wide range of activities, including filling, dredging, draining, and excavating in wetlands. Some states have partially assumed the Section 404 permitting authority where the U.S. Army Corps or Engineers (Corps) has issued a general permit for certain classes of activities—as it has, for example, with respect to Maryland's nontidal wetlands program. Other states, such as Vermont, do not have a wetlands permit program but protect wetlands through "conditional use approvals" under land use laws.

Many states specify mitigation requirements as part of their regulatory program. For example, Maryland requires applicants to "take all necessary steps to first avoid significant impairment and then minimize losses of nontidal wetlands." New Jersey requires that every freshwater wetlands permit contain a condition ensuring that "all appropriate measures have been carried out to mitigate adverse environmental impacts; restore vegetation, habitats, and land and water features; prevent sedimentation and erosion, minimize the area of freshwater wetland disturbance, and insure compliance with the Federal [Clean Water] Act and implementing regulations."

A number of states have statutes expressly authorizing wetland mitigation banks.

California

Cal. Pub. Res. Code § 30233; Cal. Fish & Game Code §§ 1775-1793; See also AB 1811 (1991) (1991 Stat. Ch. 851) re: state-owned restoration sites.

Colorado

Colo. Rev. State. §§ 37-85.5-101 to -111.

Florida

Fla. Stat. § 373.4135

Louisiana

La. Rev. Stat. Ann. § 49-214.41.

Maryland

Md. Nat. Res. Art. § 8-1209.1 et seq.

New Jersey

N.J. Stat. Ann. §§ 13.913-13 to -15.

North Dakota

N.D. Cent. Code § 61-32-05.

Oregon

Or. Rev. Stat. §§ 196.600 to .665.

Texas

1991 Tex. Sess. Law Serv. ch. 3 §§ 6.01-6.07.

Wyoming

Wyo. Stat. §§ 35-11-310 to -311.

Others have explicitly addressed banking through agency guidance or regulation. By 1993, 13 states had promulgated statutes, regulations or guidelines governing wetland mitigation banking.[1] Oregon, for example, enacted its Wetlands Mitigation Bank Act in 1987. Under this statute, mitigation banks must be publicly owned and operated, be approved by the Division of State Lands (DSL), and meet a number of criteria. Credits can only be used for mitigation of permit actions within the same "tributary, reach, or subbasin" covered by the mitigation bank and may not be used until DSL has certified them. The price of any mitigation credit must include all of the costs incurred by the state in setting up and maintaining the bank. Florida legislation in 1993 officially authorized wetland mitigation banking, although banking was already underway in the state.

Florida's regulatory framework consists of a statewide program at the Florida Department of Environmental Protection (DEP) and regional programs at five water management districts (Districts), which jointly encompass the entire state.[2] The statewide program has had permitting jurisdiction over contiguous wetlands since the late 1960s.[3] The Districts were given statutory authority to regulate dredging and filling of isolated wetlands in 1986.

Until the enactment of the Warren S. Henderson Wetlands Act of 1983, Florida's laws made no reference to the use of mitigation in the dredge-and-fill permitting process. For example, the Florida Air and Water Pollution Control Act of 1967 established public policies to protect, maintain, and improve the quality of waters for fish, wildlife, and other aquatic life, as well as for domestic, agricultural, industrial, recreational, and other beneficial uses. The act provided that no wastes could be discharged into waters without first being treated to protect beneficial uses. Wastes included point and nonpoint sources, as well as fill material. Under the act, projects were authorized without consideration of their impact on habitat or primary productivity. Generally, creation and enhancement of wetlands occurred as specific public restoration projects or use for stormwater treatment and management but rarely as mitigation. In addition, wetlands under dredge-and-fill jurisdiction comprised only a minor subset of all wetlands.

The Warren S. Henderson Wetlands Act increased the scope of dredge-and-fill jurisdiction from open-water bodies and the deepest wetlands to include most other wetlands as well. In addition, adverse impacts to habitat were included in a seven-item public interest balancing test, and mitigation was included as a tool in the permitting process.

In 1989, rules governing the use of mitigation were promulgated by DEP. The rulemaking process began in 1986, with a scheduled completion date of early 1987. However, a change in gubernatorial administration brought a new viewpoint to the rulemaking process and the rules were filed for adoption in 1988. The rules were challenged by several parties, and final adoption of the rules was delayed until January 1989. Mitigation banking was not included specifically in the rules. Rather, DEP decided that since mitigation banking was new to Florida, rulemaking should be delayed until some banks had been permitted and guidelines developed. Provisions allowing preconstruction mitigation were included in the rules to allow the use for mitigation banking.

The network of five water management districts was formed in 1972 under the Florida Water Resources Act of 1972. The act gave the Districts the authority to regulate all artificial or natural structures or constructions that affect surface water quantity or quality. It put forth a comprehensive approach to water use and management through planning, regulation, and resource management. In 1986, the Districts were given further statutory authority to regulate impacts to isolated wetlands and passed new rules that included mitigation for such impacts.

In the late 1980s and early 1990s, the state program delegated the authority for most of its dredge-and-fill permitting to the Districts with existing programs.[4] In 1993, the Florida legislature passed the Florida Environmental Reorganization Act that combined the state and regional wetlands permitting programs. The new rules written to codify this statute are currently in litigation but are expected to come through relatively unscathed. The Districts will conduct most of the permitting.[5] Included in the legislation was a requirement that the agencies implement new rules on mitigation banking. These rules are now in effect statewide.[6]

Generally, the rules are administrative in nature. Standards for determining when mitigation is needed and how it is evaluated still reside in the agencies' other rules. The banking rules promote restoration of landscapes as mitigation, provide guidelines for deciding whether a proposed bank site is likely to be hydrologically and ecologically viable in the long term, provide criteria to determine the ecological value of the bank, provide criteria regarding how many credits to assign and how to schedule their release, require that banks be preserved in perpetuity, and require separate financial mechanisms for the implementation and long-term management phases. The rules allow any type of entity to become a banker as long as it can meet the financial and land conveyance provisions of the rule, in addition to providing good mitigation. One of the issues that remains somewhat contentious is the financial responsibility requirements. The legislation required that nongovernmental banks must provide proof of financial responsibility. The rules were adopted including financial provisions for all bankers, with the most stringent requirements being for nongovernmental entities, relaxed standards for other governmental entities, and standards only for long-term management for the Districts and DEP.

Effectiveness of Mitigation

In 1990, DEP was directed to evaluate the effectiveness of wetlands mitigation that was permitted by the state. In its review, the agency found some problems with the program.[7] One-third of the required mitigation projects were never implemented, although the permitted destruction of wetlands already had occurred. Of those that were completed, few were in compliance with important aspects of the agreed-upon mitigation plan, and only 27 percent were deemed ecologically successful, with an expectation that remedial action could raise that level to 63 percent. Interestingly, these percentages are remarkably similar to those identified in a smaller study conducted by DEP in 1986 prior to drafting the mitigation rule.[8] Studies by the South Florida Water Management District (SFWMD) and the St. Johns River Water Management District (SJRWMD) found similar results (Table 4.1).

Much of the observed failure of mitigation to achieve the intended results was due to a heavy reliance on attempts to create, rather than restore, wetlands. In general, it is far more difficult to create wetlands where none existed before than to enhance or restore an existing, albeit altered, wetland. Further, many mitigation designs were evaluated without regard to the hydrologic regime that likely would result at the site. This narrow scope of review doomed many mitigation projects on the drawing board. As a result of the poor track record of wetlands mitigation projects in Florida, DEP placed greater emphasis on enhancement of degraded wetlands and restoration of historic wetlands, in conjunction with preservation (mirroring the 1990 Corps/Environmental Protection Agency (EPA) Memorandum of Agreement (MOA), as the preferred avenues for mitigation of project impacts. While these types of mitigation can be achieved on-site in some cases, they are most attainable on sites set aside for that purpose as mitigation banks. Replacement of "traditional," on-site mitigation with mitigation banks is expected to alleviate the problems identified in the reports on mitigation effectiveness by improving compliance with regulatory requirements and the performance of created, enhanced, or restored wetlands.

In 1992, a Mitigation Bank Task Force was formed by DEP, with representatives from the department, water management districts, and both environmental and development interests. The task force recommended that a number of factors be considered when evaluating banking proposals and that regulations be developed to implement mitigation banking. Among the factors recommended were that wetland restoration be preferred over wetland creation and that banks be managed over the long term. It suggested that DEP hold off on developing regulations until after the department had more experience overseeing mitigation banking.

In 1993, the Florida legislature agreed with the task force and determined that mitigation banks can minimize mitigation uncertainty and provide ecological benefits.[9] Accordingly, the DEP and regional water management districts were directed to adopt rules governing establish-

Table 4.1 Success of Mitigation Projects in Florida

Study[a]	Mitigation Projects Implemented (%)	Ecological Success: Successful (%)	Ecological Success: Potentially Successful[b] (%)
Walker, 1986[c]			
Saltwater	> 79	38	79
Freshwater	> 50	12	50
Combined	> 72	31	72
Beever, 1990[d]	—	58	19
DER, 1991			
Saltwater	—	45	> 90
Freshwater	—	12	41
Combined	66	27	63
Erwin, 1991[e]	40	10	40
Lowe, 1992			
Saltwater	—	100	100
Freshwater	—	10	82
Combined	—	59	83

[a]The number of mitigation sites reviewed by each study were Walker, 32; Beever, 43; DER, 119; Erwin, 40; and Lowe, 326.
[b]Would likely become successful if remedial actions were implemented.
[c]The estimate of implemented mitigation for Walker (1986) is based on a category of failed mitigation, which includes mitigation that was not implemented.
[d]All sites for Beever (1990) were saltwater.
[e]All sites for Erwin (1991) were freshwater.

ment and operation of mitigation banks pursuant to several legislative rulemaking directives.[10] These directives established core policy for mitigation banks,[11] such as authorization for both public and private banks; a requirement that bank credits be based on net ecological value added by mitigation bank activity; provision for enhanced credit incentives for achieving mitigation success prior to credit withdrawal; ability for bankers to sell credits to third parties; and potential recognition of creation, restoration, enhancement, and preservation activities for both wetlands and uplands as bank credits.

The DEP and regional water management districts adopted substantially similar rules in 1994.[12] The rules provide criteria for creation and use of banks to offset adverse impacts of activities regulated under state wetlands and management and storage of surface-waters permit programs.[13] They do not supersede or modify otherwise applicable permit review criteria or the requirements of any other rules, such as delineation of wetlands and mitigation ratios.[14]

One of the reasons why many mitigation projects fail is the inability of regulatory agencies to consider adequately the long-term ecological and land use conditions within which particular sites exist. Too often regulatory agencies have been caught in the project-by-project syndrome and made decisions that do not provide for long-term sustainability of isolated or regional systems. Therefore, in order to derive the greatest benefit from a restoration activity, a mitigation bank should be designed in an ecosystem management context.

Florida is moving toward the concept of ecosystem management in its planning, land acquisition, and regulatory arenas. DEP defines ecosystem management as "an integrated, flexible approach to management of Florida's biological and physical environments—conducted through the use of tools such as planning, land acquisition, environmental education, regulation, and pollution prevention—designed to maintain, protect, and improve the state's natural, managed, and human communities."[15] Watersheds are an appropriate unit in which to consider ecosystem management.

Watershed planning is derived from many of the same principles as ecosystem management. It is predicated upon the understanding that the actions of each component have an impact on all parts. Yet, balancing sometimes competing goals of providing or enhancing water supply, flood protection, and environment functions of the basin is complicated by the individual actions of a myriad of governmental entities with interest over the land. Nevertheless, to provide for sustainable natural systems, agencies with broader jurisdictions must undertake a comprehensive approach to resource management activities. As discussed previously, one of the criteria for approval of mitigation banks is an increase in the ecological value of the regional watershed due to the mitigation proposed at the bank.

Implementation of regional ecosystem management or restoration goals requires a great deal of coordination among counties or municipalities with interest in portions of the watershed. Thus, a mechanism for integrating urban land use decisions at the local level and ecosystem management goals at the regional or watershed level must be developed.

Various states have enacted growth management laws to provide such a mechanism. These laws typically require each county in the state to prepare a growth management plan that includes policies and objectives to monitor and accommodate growth while maintaining or enhancing the level of urban services. These plans also include provisions for the protection of natural resources. In Florida, for example, each county prepares a plan that is revised every five years. Each plan is reviewed by all applicable agencies of the state to ensure consistency with statewide mandated minimum standards, goals, and policies prior to its final adoption.

In Dade County, Florida, the private sector has realized the benefits of participating in a process, such as a SAMP or ADID, that will reduce the uncertainty of the permitting process. Most developers lack interest in building and managing natural areas and would much rather pay to have another entity, such as a mitigation bank, perform those functions. Such regional planning tools have improved the permitting process in Dade County, while providing regionally significant and functional mitigation. The designation of certain areas as unsuitable for development has relieved development pressure in some environmentally sensitive areas.

What Constitutes a Bank?

Before 1993, mitigation banks in Florida were constructed for a permittee's own use, to offset future adverse impacts. Most mitigation banks were established by local governments or by developers of large projects. Entrepreneurial banks were not considered to be acceptable because there was no link between offsetting impacts and the permitted losses. DEP believed that those responsible for the adverse impacts to wetlands should be responsible for mitigating the loss of resources. Since 1993, however, a variety of banking options exist. The agencies' mitigation rules and the new combined rules on mitigation allow permittees to contribute funds to ongoing environmental restoration projects where such contributions would cover the cost of work that would mitigate a permittee's impacts. Under the new rules, anyone can create a mitigation bank and sell credits to the open market. The free market will prevail. The cost of credits is not regulated by the agencies, nor is the speculative sale of credits. Contracts with third parties (bankers) to develop and implement banks may become common if consortiums of regulated interests work toward that end.

Because the environmental standards for mitigation reside in existing rules, the new banking rules focus on administrative and evaluative details. New evaluative processes exist regarding how to consider the likelihood of the chosen bank site to offset the impacts.

One type of partnership that is gaining strength is private development of a mitigation bank on public lands. A few such banks have been implemented or are in final planning stages. For example, one of the first entrepreneurial mitigation banks permitted by the Corps is in Pembroke Pines, Florida, about 20 miles southwest of Fort Lauderdale. The 350-acre site, owned by the city of Pembroke Pines, is choked with exotic plants that have severely limited its value to native wildlife. The Florida

Wetlandsbank is restoring the degraded wetlands to create a mix of habitat types typical of the Everglades, including cypress stands, emergent marshes, sawgrass prairie, wading bird–feeding areas, and tree islands (see case study 2 following the chapters). In return, the city will receive an improved wetlands/park with nature trails and picnic areas.

Under a licensing agreement with the city, Florida Wetlandsbank will restore the wetland, sell mitigation credits to developers, and, after maintaining the restored wetland for five years, turn responsibility for maintaining the site back to the city. The bank is responsible for designing, permitting, and constructing the ecosystem and pays the city $1,000 per acre for an escrow fund to maintain the site, plus another $8,800 for a performance bond to guarantee that the work is completed satisfactorily.

The use of public lands for private banks raises several issues. On one hand, it provides a mechanism to fund restoration of public lands that otherwise would remain unrestored: private entities will have personnel and funding to create the bank and assume legal and fiscal liabilities. On the other hand, it may provide unfair economic advantage over competing banks on private land, and there may be less control over the quality of the restoration.

In order to obtain a mitigation bank permit, a series of criteria must be met (Section 62-342.400 FAC).

1. The banker shall provide reasonable assurance that the proposed Mitigation Bank will:
   - (a) improve ecological conditions of the regional watershed;
   - (b) provide viable and sustainable ecological and hydrological functions for the proposed mitigation service area;
   - (c) be effectively managed in the long term;
   - (d) not destroy areas with high ecological value;
   - (e) achieve mitigation success; and
   - (f) be adjacent to lands which will not adversely affect the long-term viability of the Mitigation Bank due to unsuitable land uses or conditions.
2. A Mitigation Bank may be implemented in phases if each phase independently meets the requirements of subsection 62-342.400(1) above.
3. The banker shall:
   - (a) have sufficient legal or equitable interest in the property to meet the requirements of section 62-342.650; and
   - (b) meet the financial responsibility requirements of section 62-342.700.

The criteria provide guidance for mitigation banking proposals to ensure that the mitigation proposed at the bank will improve the ecological

value of the regional watershed within which the bank resides and to increase the likelihood that the bank will be viable over the long term and relatively self-maintaining. A banker also must have the right to use the land in perpetuity and be able to provide financial assurance mechanisms for the implementation and long-term management of the bank. As stated in the preamble to the legislation on mitigation banking, the state expected that mitigation banking should yield better long-term results than was experienced at the time.

A mitigation bank permit authorizes implementation and operation of a bank and establishes rights and responsibilities of the banker and the permit agency for implementation, management, maintenance and operation of a bank.[16] The permit must include[17]

1. a description of the mitigation service area;
2. a maximum number of mitigation credits available for use when the mitigation bank or phase is deemed successful, type of mitigation credits awarded, and the number and schedule of mitigation credits available for use prior to success;
3. the success criteria by which a mitigation bank will be evaluated;
4. financial responsibility mechanism(s) which must be employed by the banker, including the procedure for drawing on the financial mechanisms by the permit agencies, and provision for adjustment of the financial responsibility mechanism;
5. requirements for executing and recording a conservation easement or conveyance of the fee interest for mitigation bank land;
6. a ledger listing available mitigation credits;
7. a schedule for implementing the bank, and any phases; and
8. long-term management requirements.

A bank permit automatically expires five years from the date of issuance if the banker does not record a conservation easement or convey a fee simple interest, as appropriate, over the real property within a bank, or phase, in accordance with a permit requirement.[18] If a property interest is not required to be recorded, a permit will automatically expire if construction is not commenced pursuant to the construction schedule established by a permit.[19] With these exceptions, a bank permit is perpetual unless revoked or modified.[20]

Many of the items covered in the banking rule were included to improve mitigation practices. Mitigation banks must be established, although not necessarily completed, prior to the sale of mitigation credits. Specified percentages of the total credits can be released in phases based on the amount of work completed at the bank. If a bank ever is not in compliance with a mitigation bank permit, the sale of credits will be frozen until the noncompliance is remedied. Financial assurance mechanisms for both implementation and long-term management must be fully funded prior to sale of credits. The permitting agency must be the beneficiary of the financial assurance so that it may complete the work should the banker default on its obligation.

Where feasible, banking will be used in conjunction with on-site mitigation. Thus, by concentrating restoration and enhancement of wetlands into fewer sites, the agencies' staff will be able to inspect sites adequately for compliance.

One of the issues the new rules had to address was how to weigh withdrawal of credits from a bank located within, rather than outside of, a regional watershed. This reference to "regional watershed" created a new standard for the geographic scale whereby mitigation for impacts may be considered acceptable. Previously, mitigation usually had been limited to a development site, even though this frequently resulted in unsuccessful mitigation. The Districts developed maps of the state's regional watersheds as a means of defining this term, although the size of such watersheds varies among Districts. Two Districts (NWFWMD and SRWMD) experiencing minor development pressures specified large watershed boundaries, as did the SWFWMD, while the two more rapidly developing Districts (SJRWMD and SFWMD) specified much smaller watersheds.

One of the more difficult issues associated with banking has been determining the appropriate geographical boundary within which credits can be sold for a given bank. This geographic area is referred to as the Mitigation Service Area (MSA). DEP has been using a watershed-based approach, as have the Districts, while local governments have a tendency to prefer that mitigation occur in the same county as the impacts. These purely political boundaries conflict with watershed-based decision making. Another portion of the legislation that required the new rules on mitigation banking states that where local and state/regional mitigation requirements conflict, the state/regional mitigation determination will prevail. This language has not yet been challenged legally. A concern of the state/regional agencies is that mitigation areas in urbanizing counties may be less likely to provide the long-term habitat functions of banks located in areas of the watershed with less development pressure.

Currently, DEP is working toward the statutory directive to encourage and participate in mitigation banking. First, the agency is working with the Florida Department of Transportation to locate mitigation bank sites for its mitigation needs. Second, DEP is working with all levels of government to develop jointly, by 1996, a restoration project and mitigation bank-siting plan for the SFWMD, which covers nearly one-third of the state. This approach will be adopted by the remaining districts as well. Third, while DEP will not develop banks on its own land, it will continue to allow mitigation banks to be developed on its lands. While no banks that use state-owned land have been permitted yet, three sites are currently being considered by third parties as possible banks sites. Third-party banks would need to conform with DEP-identified, priority restoration needs, as listed in the statewide identification of restoration and mitigation bank sites. Since the late 1980s, the department has allowed permittees to contribute to identified or ongoing restoration projects on its lands. Usually, the permittee develops a restoration plan in consultation with department staff, and then carries out the work with department oversight. In one project, the Florida Department of Transportation filled 19.4 acres of ditches, which restored sheet flow to 9,600 acres of state-owned wetlands at virtually no cost to the department. In other cases, the mitigation analysis determined how much mitigation or restoration was necessary and the permittee was required to contribute enough money to pay for that acreage of restoration on state lands. Of course, the restoration had to be in-kind and proximate to the impact site.

The water management districts are approaching banking in a variety of ways. The SJRWMD and SFWMD are planning to establish banks on their lands in the next two years. The SRWMD and NWFWMD, however, do not have the resources to develop mitigation banks on their lands but will identify restoration sites for project-by-project mitigation. All of the Districts will participate in the statewide restoration site planning efforts that are just beginning. At present, permittees can contribute to ongoing restoration projects on district lands.

Interagency Cooperation and Conflicts

In order to develop a public or private mitigation bank, a bank developer must identify and include, in each step of development, all of the agencies from which authorizations will be required. The identification of the role of each party will promote the timely development and implementation of the project. Critical to this process is the identification of an entity or lead agency who will assume the responsibility for collecting and distributing information, as well as facilitating all meetings.

For private mitigation banks, the entrepreneur will perform all of the necessary tasks to implement the plan. Under such circumstances, agency participation is commonly limited to oversight of the project. For public banks, the designation of certain tasks to agencies having particular expertise often results in a sense of proprietorship on behalf of the participants ("our" bank versus "your" bank) and helps agencies share the expense of developing an effective plan.

Often regulatory decisions are made without necessary information because of pressures placed on agencies to render quick decisions or because they do not have adequate resources to perform the desired studies. Here is where intergovernmental cooperation can provide the greatest benefit. By studying and evaluating carefully all the variables that could influence the effectiveness of the restoration activity, the probability of success is enhanced greatly.

The regulation of development in wetlands has become so complex that even projects undertaken for the exclusive purpose of restoring or enhancing the productivity of natural areas often are left at the drawing table, a victim of interagency conflict. Moreover, conflicts arising from institutional mistrust and varying regulatory or land management philosophies often present obstacles in the development of a mitigation bank. Such problems, however, can be overcome with a clear and agreed upon set of goals. Issues that typically generate agency discord include techniques for assessing present and future habitat function, on-site (within basin) versus off-site (outside basin) mitigation, in-kind versus out-of-kind mitigation, and the amount of mitigation credit given for certain mitigation activities. The use of government property for mitigation rather than private lands also has generated controversy in the early history of mitigation banking.

Two mitigation banks established in south Florida in 1995 deal with yet another facet of this issue, the use of public land by a government entity (Hole-in-the-Donut in Everglades National Park) versus the use of public lands by a private, for-profit venture, such as the Florida Wetlandsbank in Broward County. Both banks seek to restore wetlands. No public source of funding is currently available, or is likely to become so, to undertake this work. In both cases restoration and active management are desperately needed to prevent the continued expansion of exotic vegetation on the sites and to eliminate a seed source for further degradation of surrounding native communities.

Opponents argue that the establishment of mitigation banks on public land represents an inefficient use of limited environmental funding. Others argue that the agencies should not accept "dirty money" or funds generated by the loss of wetlands.

Yet, development will continue to affect wetlands regardless of where the mitigation is undertaken. Additionally, when federal, state, and local park systems' budgets consistently are cut in favor of other critical services, other sources of funding must be found to maintain natural systems.

One of the more difficult issues associated with banking has been arriving at a consensus on the appropriate geographical boundary within which credits can be sold for a given bank. This geographic area is referred to as the Mitigation Service Area (MSA). DEP has been using a watershed-based approach, as have the Districts. Local governments have a tendency to prefer that mitigation occur in the same county as the impacts. These purely political boundaries conflict with watershed-based decision making. Another portion of the legislation that required the new rules on mitigation banking states that where local and state/regional mitigation requirements conflict that the state/regional mitigation determination will prevail. This language has not yet been tested. A concern of the state/regional agencies is that mitigation areas in urbanizing counties may be less likely to provide the long-term habitat functions of banks that are located in areas of the watershed with less development pressure.

Credit Assessment and Bank Monitoring

Under Florida's new rules, credits for mitigation can be traded prior to initiation of the restoration activities or in various stages of implementation. The rules allow the release of credits when the bank site is either placed into a conservation easement or deeded over to the issuing agency. A conservation easement must be granted to both the issuing agency (DEP or the District) and to the other agency (the District or DEP). This double granting of the easement provides an extra level of protection, because under Florida law conservation easements may be returned to the grantor at the grantee's discretion. Lands deeded to the issuing agency would be considered as conservation lands in perpetuity. No credits may be released until this step has occurred. Credits may then be released in phases, however, the agencies require that a percentage of the credits be withheld until the mitigation meets the criteria for success, as determined by the agency. Criteria for success are based on vegetative, hydrologic, and soils parameters. For private or entrepreneurial mitigation banks this means that credits will be released at a rate sufficient to get the bank started, with the majority of credits being made available at periodic points during implementation. When private entities use public land for a mitigation bank, and the land is already preserved, the first credits would be released periodically as development of the bank proceeds.

Originally, it was thought that banks should be declared successful and complete prior to withdrawal of credits. In theory, this would reduce the risk of failure and help ensure that there is "no net loss" of habitat function. It would also hold the banker to a higher mitigation standard than

DEP requires for traditional, on-site mitigation. DEP believes that the incentive of allowing credit sales early in the banking process, coupled with holding out a portion of credits until success is achieved, is a more reasonable standard.

In some cases, the agencies collaborate to determine the amount of credits available and delineate the development area or types of projects that may use the bank. A single agency is then identified, in a Memorandum of Understanding, permit, or similar legal vehicle, to undertake all ledger accounting, compliance, and enforcement responsibilities.

Assurances and Certainty

When a private bank sells credits prior to the implementation of the work, the entity undertaking the restoration is required by the new rules to provide some tangible, liquid form of financial assurance, such as a letter of credit or a performance bond, to ensure compliance. The permit agencies have not yet adopted model forms for providing proof of financial responsibility. Accordingly, bank applicants must provide initial drafts of financial responsibility instruments for agency approval.[21]

Financial responsibility mechanisms must be established with a state or national bank, savings and loan association, or other financial institution licensed in the state.[22] Letters of credit may be used and if so, must be issued by an authorized entity whose letter of credit operations are regulated and examined by a federal or state agency.[23] Surety bonds may be employed provided they are issued by a surety company registered with the state.

Financial responsibility mechanisms cannot expire or terminate prior to completion of all applicable bank permit conditions, and they must be effective at least 60 days prior to start of construction of bank phase or as otherwise required by a permit.[24]

If a bank fails to comply with permit conditions, the permit agency, upon reasonable notice to the banker, may draw upon the financial responsibility instrument. Thus, the financial assurance mechanism should, at a minimum, be equal to 100 percent of the total cost of the restoration including a reasonable surcharge or overhead amount to cover the expense of having a government entity contract for the supervision of construction. Proof of financial responsibility is not required where the construction and implementation of a bank, or a phase, is completed and deemed successful prior to the withdrawal of any credits.[25] Because the long-term health of the bank is so important, Florida's new rules require that a separate long-term management trust fund be established. The fund is to be perpetual and provide sufficient interest income to pay for ongoing, routine maintenance of the bank site. All costs associated with the long-term management are estimated to determine the amount needed in the fund. To determine the amount of financial responsibility required for construction and implementation and long-term management, a banker must submit a detailed written estimate, in current dollars, of the total cost of construction and implementation and long-term management.[26]

The cost estimate for construction and implementation must include all costs associated with earthmoving, structure installation, consultant fees, monitoring activities, and reports.[27]

Long-term management cost estimates must include costs to maintain and operate any structures, control nuisance or exotic species, support fire management, pay consultant fees, monitor activities and reports, and meet any other costs associated with long-term management.[28]

Cost estimates with verifiable documentation must be submitted to the permit agency for approval. Estimates must be based on a third party performing the work at fair market value.[29] The source of cost estimates must be indicated.[30]

Because of the effect inflation can have on costs, provisions for periodic adjustment of the cost estimates for the construction and implementation stage and the long-term management stage are critical. For this reason there are mandatory adjustments of the cost estimates every two years. The cost adjustment must be submitted to the permit agency for approval accompanied by supporting documentation. Additionally, a permit agency may require adjustment of the amount of financial responsibility if such adjustment is required in order to meet permit conditions.

Some flexibility in the rules is necessary, because government entities are less likely to be able to provide financial assurance. Mitigation for public banks commonly is in the form of a contribution as a part of a "pay as you go" scenario. Even though a bank undertaken by a public entity typically will not be required to post a bond, the agency promoting the mitigation bank should be held liable for the proper implementation of the restoration activity, via an MOU or similar agreement. When executing MOUs for the implementation of a mitigation bank, the agencies should seek to delineate clearly the responsibility of all parties and to streamline their respective regulatory processes. At a minimum, the agreement should contain clear goals, a statement of work including all accounting, compliance, and enforcement responsibilities, a schedule for providing periodic reports or updates, a conflict resolution mechanism, a description of the legal authority of each party to enter into the agreement, and default provisions.

If possible, all other regulatory requirements within the area to be developed or serviced by the mitigation bank should be reconciled to produce a single set of criteria. This in turn would permit the delegation of regulatory authority to the lowest level of government technically capable of assuming such delegation. For the federal government, this can be accomplished through the issuance of a General Permit for the limited geographical area. Assumption of state delegation by a local government may be in the form of a contract or agreement.

The results of various mitigation studies have shown that the major problems with mitigation were caused by design flaws and permit noncompliance. This finding has caused the agencies to place greater emphasis on mitigation that is conducted prior to or concurrent with the permitted impacts. In addition, the agencies shifted their emphasis away from wetland creation to wetland restoration and enhancement.

In early mitigation bank permits, DEP assigned the availability of mitigation credits based on likelihood of success of created, enhanced, or restored wetlands. For projects likely to be successful, such as creation of a salt marsh or rehydration of a freshwater forested wetland, credits were made available upon completion of the created or enhanced wetland. Credits were released at a ratio consistent with concurrent mitigation. As time progressed and the mitigation came closer to success, the mitigation ratios became more favorable to the permittee, as defined in the bank permit. For more difficult types of mitigation, such as creating a freshwater forested wetland, mitigation credits were not allowed to be released until most of the success criteria were met. Once the mitigation was deemed successful, a more favorable ratio was allowed for credit release.

Under Florida's Mitigation Bank Rule, credits are released in phases, as specified in the permit. These phases of release will be evaluated on a project-by-project basis. If the project is in compliance with the terms of the permit, then the agency must release the credits as scheduled. Therefore, under these rules it is imperative to evaluate the bank proposal to ensure that there is sufficient reasonable assurance that the mitigation will become ecologically successful. It is equally important to write a clear permit that addresses the ecological success of the project.

In Florida, mitigation projects will be declared "successful" when it *appears* that the created, restored, or enhanced wetland will continue to grow and develop into the type of ecological community that was intended. In making such a determination, the agencies attempt to balance the cost of monitoring for the permittee with the likelihood that the losses will be mitigated. Many projects will require ongoing, but low intensity, maintenance. Thus, in Florida, bankers must provide assurance that if they default on their long-term maintenance obligation, the agencies will have recourse to fund the necessary maintenance of the site. As Florida's rules on mitigation banking require that prospective bankers establish an acceptable long-term management fund before credits are made available for sale, the challenge to regulators is to determine the level of assurance needed to ensure a project will be successful without overly hindering investment in the creation and operation of mitigation banks.

Long-term management of banks is critical. It will be the responsibility of the banker to ensure that routine management be undertaken. Most likely, this phase of the project will be turned over to an entity that specializes in conservation land management. For public lands, the landowner likely will become the manager, even when the bank was developed by a private entity. For private banks, conservation land stewards, such as The Nature Conservancy, are likely to become long-term managers. Professional land managers who currently manage for silviculture, rangeland, hunting, or similar activities are likely to expand their expertise to include management for conservation of ecosystems.

As with the construction and implementation phase of mitigation banks in Florida, financial assurance of the long-term management phase is required in the form of a trust fund agreement. All costs related to the long-term management are required to be estimated for the purpose of determining the amount with which the fund must be endowed. To account for the effect of inflation and improvements in land management technology, the amount must be adjusted every two years, based on a reestimation of costs. The fund may then be adjusted up or down, depending on the outcome of the new estimate.

Most banks probably will be created by restoring historic hydrologic conditions of a wetland followed by replanting and fire management. This type of mitigation bank will be relatively simple and inexpensive to maintain over the long term because the site will be returned to its "natural" state. Because of the extensive infestation of exotic plants in south Florida, however, more intensive forms of management will be necessary. Given the scale of infestation in south Florida, it is unlikely that a site could become free of exotic plants for long. Any banker proposing a bank that will require removal of exotic plants must provide extra assurances that the eradication of such plants will be successful.

Conclusion

During the development of Florida's wetland mitigation banking rule, one of the greatest concerns of the environmental community was that banking would allow developers to circumvent the requirements to avoid or minimize adverse impacts to wetlands. However, the mitigation sequencing requirement remains central to both state and federal wetland regulatory programs. The existence of a mitigation bank should have no bearing on regulatory decisions about whether or not a particular development in wetlands should be permitted. Rather, once the agencies determine that a wetland impact is permittable (after avoidance and minimization), then the use of credits from a bank simply becomes one of the potential mitigation options an applicant may propose to offset the impact.

In addition, regulators retain the responsibility to determine, on a permit-by-permit basis, whether to permit the use of credits from a mitigation bank to offset the proposed wetland impacts. Florida's rules specify that the use of a mitigation bank is an appropriate, desirable, and permittable mitigation option when a bank offsets the adverse impacts of the project and (a) on-site mitigation opportunities are not expected to have comparable long-term viability due to factors such as unsuitable hydrologic conditions or ecologically incompatible adjacent land uses or (b) use of the bank would provide greater improvement in ecological value than on-site mitigation. Similarly, federal guidelines ("Establishment and Use of Wetland Mitigation Banks in the Clean Water Act Section 404 Regulatory Program" signed August 23, 1993) regarding the use of mitigation banks explain that the Corps and EPA's preference for on-site, in-kind mitigation does not preclude the use of mitigation banks. These agencies have agreed that mitigation requirements may be satisfied through the use of mitigation banks, provided their use is consistent with standard practices for evaluating mitigation proposals outlined in the "Memorandum of Agreement Concerning the Determination of Mitigation under the Section 404(b)(1) Guidelines" signed by the Corps and EPA on February 6, 1990. In determining whether use of a particular mitigation bank is appropriate for offsetting impacts, the federal agencies consider wetland functions, landscape position, and affected species populations at both the development site and the mitigation bank site. Federal guidelines further specify that mitigation for wetland impacts should occur, where appropriate and practicable, within the same watershed as wetlands being affected by development.

For permittees, perhaps one of the most appealing aspects of buying credits from a mitigation bank to offset wetland impacts is the transfer of mitigation responsibility to a bank operator or owner. For traditional on-site mitigation practices, a permittee is held responsible for implementing the permitted mitigation plan successfully. The costs and time associated with carrying out the mitigation can be substantial, especially given the likelihood of mitigation failure, the need for remedial measures, and the potential for the project to be subject to noncompliance or enforcement action.

One of the greatest criticisms of existing regulatory programs is that developers often fail to implement the required compensatory mitigation, or if they do, the mitigated wetlands fail anyway due to poor design, implementation, and maintenance. The advantages of mitigation banks is that they can consolidate numerous small, isolated wetland mitigation projects into one large block of created, restored, or enhanced wetlands, providing greater assurance that the restoration will be accomplished and will work as planned.

The larger blocks of wetlands are much easier to evaluate and determine whether the required mitigation has been completed correctly. In addition, mitigation banks provide the opportunity to locate restored or created wetlands in areas of greatest need in the watershed. For example, there are many streams with valuable wetland habitat in areas both upstream and downstream of degraded or drained wetlands. These drained wetlands were once part of the stream's aquatic ecosystem. Such areas could be restored to provide not only the values of the restored wetlands, but to connect the other two existing valuable wetland areas as well. In this way, reestablishing wetland corridors could increase the value of the wetlands restoration beyond the value of the individual wetlands being restored. Such corridors help prevent certain species from being isolated from the ecosystem by allowing wildlife to move freely from one aquatic area to another.

Mitigation banking can also improve the rate of success of wetlands mitigation by providing a more favorable location for created or restored wetlands. For example, wetlands restored, created, enhanced, or preserved in undeveloped or sparsely developed areas will have a much greater chance of survival than wetlands established or preserved in highly developed areas.

Increasingly, mitigation banking will play a greater role in regulatory programs. The agencies view mitigation banks in a very positive way due to the ecological and administrative benefits of mitigation banking as a method of providing compensatory mitigation.

Notes

1. These states include California, Colorado, Louisiana, Maine, Maryland, Minnesota, New Hampshire, New Jersey, North Dakota, Oregon, Texas, Wisconsin, and Wyoming.

2. These are the St. Johns River Water Management District (SJRWMD), the Southwest Florida Water Management District (SWFWMD), the South Florida Water Management District (SFWMD), the Suwannee River Water Management District (SRWMD), and the Northwest Florida Water Management District (NWFWMD).
3. "Contiguous" wetlands are those that can be traced from a stream or estuary to its landward edge. They are not wetlands that are isolated from a stream or estuary by uplands.
4. These are the SJRWMD, the SWFWMD, and the SFWMD.
5. In addition to the three districts currently having the delegation of permitting, the SRWMD will start up a program. The fifth district, NWFWMD, will not have a program due to lack of funding and all permitting will be done by the department.
6. DEP: Florida Administrative Code (FAC) Rule 62-342; SJRWMD: FAC Rule 40C-4, Applicant's Handbook, Part II, Section 16.1.6; SWFWMD: FAC Rule 40D-4, Basis of Review, Appendix 6; and SFWMD: FAC Rule 40E-4, Basis of Review, Part II, Appendix 8. Mitigation banks within the bounds of the SRWMD and NWFWMD will be handled by the DEP.
7. Florida Department of Environmental Regulation, *Report on the Effectiveness of Permitted Mitigation*. Tallahassee, FL, 1991.
8. Walker, S., "State Regulatory Standards for Mitigation and Restoration," *Mitigation of Impacts and Losses*. J. A. Costlier, M. L. Quammen, and G. Brooks, eds., pp. 79–81, 1988.
9. Fla. Stat. § 373.4135 (1993) and Fla. Admin. Code R. 17-342.100(3).
10. Id.
11. Fla. Stat. § 373.4135(1)–(11).
12. This discussion addresses the DEP rules, Fla. Admin. Code, Chapter 17-342.
13. Fla. Stat. Ch. 373, Part IV (1993).
14. Fla. Admin. Code, R. 17-342.100(2).
15. Florida Department of Environmental Protection, *Beginning Ecosystem Management*. Tallahassee, FL, 1994.
16. Fla. Admin. Code, R. 17-342.750.
17. Fla. Admin. Code, R. 17-342.750(1).
18. Fla. Admin. Code, R. 17-342.750(2).
19. Id.
20. Id.
21. Fla. Admin. Code, R. 17-342.700(2).

22. Fla. Admin. Code, R. 17-342.700(3).
23. Id.
24. Id.
25. Fla. Admin. Code, R. 17-342.700(4).
26. Fla. Admin. Code, R. 17-342.700(6).
27. Id.
28. Id.
29. Id.
30. Id.

5

Point/Counterpoint: Two Perspectives on Mitigation Banking

Both the regulatory and the environmental community seem dissatisfied with the current wetlands regulatory program. Developers claim that the 404 permit program is unpredictable, time-consuming, and expensive, while environmentalists bemoan the loss of wetlands and the poor track record of restored or created wetlands.

In theory, mitigation banking offers the predictability and convenience that developers seek while improving greatly the rate of success of wetlands mitigation. In addition, such banks could provide a reasonable solution to the problem of mitigating small, isolated wetlands, particularly if used in conjunction with watershed management plans. For these reasons, mitigation banking enjoys strong support among developers, landowners, economists, and the U.S. Army Corps of Engineers (Corps). The Environmental Protection Agency (EPA) is warming to the concept. Yet, many environmentalists remain skeptical. Among their main concerns are that developers will no longer seek to avoid filling wetlands, knowing that they can simply write a check to purchase mitigation credits and that created or restored wetlands in banks will ultimately fail, leading to a net loss of wetlands.

Given the short history of mitigation banking, many issues remain unresolved. Two opposing views on mitigation banking, that of the development and environmental communities, are presented here.

A View from the Private Sector

Virginia Albrecht and Michelle Wenzel

Introduction

At this writing, during the summer of 1995, the 104th Congress is considering amendments to the Clean Water Act (CWA). Ideas abound for reform of CWA Section 404, the federal wetlands program. One of the most provocative of these is the concept of "mitigation banking." The term "mitigation banking" is too new to be defined precisely; its boundaries are currently as fluid as those of the Mississippi River. The U.S. Army Corps of Engineers, the Environmental Protection Agency, and other federal agencies involved in wetlands regulation have proposed to carve out a channel for the term by defining it as:

> [W]etland restoration, creation, enhancement, and in exceptional circumstances, preservation undertaken expressly for the purpose of mitigating unavoidable adverse wetland losses in advance of development actions, when compensatory mitigation cannot be achieved at the development site or is not as environmentally beneficial. It typically involves the consolidation of fragmented wetland mitigation projects into one large contiguous site. Units of restored, created, enhanced or preserved wetlands are expressed as "credits" [that] may subsequently be withdrawn to offset "debits" incurred at a project development site.[1]

In this general form, mitigation banking has piqued the interest of the private sector. Landowners are continually searching for ways to reduce the risks that pervade the development business. Not surprisingly, *certainty* in the planning and implementation of development projects is viewed as a kind of Holy Grail by the industry. Any mechanism that would reduce the uncertainty of the notoriously arcane Section 404 regulatory process would be a real boon. Moreover, the private sector recognizes that the American public, committed as it is to environmental protection, wants to live and work in harmony with the natural world. Developers are therefore on a constant quest for mechanisms that will help them achieve cost-effective environmental stewardship.

Mitigation banking appears to be just the kind of dual-purpose mechanism the private sector seeks. The concept is capable of injecting certainty into the development process while simultaneously offering tangible environmental benefits. To date, mitigation banking has not been used to its full potential because of regulatory-based barriers to entry that are motivated in part by fears that mitigation banking will lead to a net loss of wetlands.[2] Properly implemented, however, mitigation banking can be advantageous for the economy and the environment alike.

Mitigation Banking Is Good Policy

For centuries, people perceived wetlands as nuisances that were to be filled and made into productive land as quickly as possible. When they thought about wetlands at all, they thought of useless swamps that served as breeding grounds for noxious creatures, as places that could not be built upon, farmed, or otherwise used in any beneficial way. Today, these ideas have been relegated to the scrap heap of discredited truisms.

We as a society have recognized, in just the last few decades, that wetlands provide a wealth of valuable functions for humans and the environment. Wetlands provide natural means for achieving flood control, erosion control, pollution assimilation, and sediment trapping, while simultaneously furnishing habitat for a myriad of wildlife and plant species and open space for human recreational purposes. These functions are considered so important that the protection of wetlands is now almost universally believed to be in the public interest.

But the public interest in conserving wetlands is not matched by an equivalent level of public control over wetlands resources. Rather, the private sector owns approximately three-quarters of all wetlands in the continental United States.[3] This pattern of ownership gives rise to policy questions that have important ramifications for mitigation banking. In our society, where the right to own and use private property is an essential element of individual liberty, how can we control the use of private property to achieve broad public goals? And how can we achieve that control when the public fisc is nearly as limited a resource as wetlands themselves?

A few premises are clear. Plainly, development *is* going to occur, growth *is* going to occur. People need jobs and places to live, schools and roads, hospitals and grocery stores, and yes, even shopping malls. Moreover, agriculture and silviculture are important to our national well-being. Future development will help provide for these needs. With the existing broad definition of wetlands, few significant projects will not affect wetlands, because wetlands are essentially everywhere. Development also affects other important resources treasured by the public but owned privately, such as endangered species habitat and open space. These premises tee up the classic wetlands conflict: the public wants to protect and restore these private lands for their wetlands values but does not have the money, or the will, to purchase the property. The landowner, on the other hand, wants to develop his land but in order to do so must satisfy rigorous environmental standards.

To date, the Section 404 regulatory program has had some success in limiting wetlands loss. However, zealous administration of the program, wherein form overwhelms function and flexibility is eschewed, has alienated landowners. Governmental programs that call for the sacrifice of private interest for a broader public good are most effective when they have the consent, or at least the respect, of the governed. But the existing regulatory program is woefully short on both. Thus, the challenge is to win back landowners' support for the wetlands program. One way to do this is to establish incentives, built on a recognition of private as well as public interests, that can inspire private investments in wetlands preservation, restoration, enhancement, and creation.

The policies underlying the Section 404 program can only benefit from thoughtful scrutiny and careful adjustment. In the wetlands context, law has lagged behind policy; it is time for Congress to ratify administrative policies or set the program on a new path. Mitigation banking can be one element of meaningful wetlands reform. Its implementation will provide a new avenue for the private sector to contribute to wetlands conservation efforts while pursuing economic growth that is also in the public interest.

Mitigation Banking Is Good for the Environment

Mitigation Banking Focuses on Environmental Results

Under the present regulatory scheme, a person who wants to build houses, grocery stores, hospitals, airports, roads, driveways, marinas, or just about any other imaginable project, large or small, in wetlands has to apply for a Section 404 permit to do so. Because the administrative definition of wetlands is so broad (an area need not ever have water at the surface to qualify as a wetland) and because federal regulation reaches all areas that meet the technical definition of wetlands, no matter how small or how isolated, there are few projects of any size that do not require a federal wetlands permit. For example, in 1994 alone the Corps issued approximately 90,000 permits under its individual and general permitting programs.[4]

The federal agencies administering the wetlands program have established a rigid sequence of requirements for applicants to follow when applying for Section 404 permits. Applicants must first avoid, to the extent possible, any adverse impacts their projects would have on wetlands, then minimize any unavoidable impacts, and finally compensate for—or mitigate—any remaining adverse impacts.[5] What this means in practice is that applicants are routinely required to replace natural wetlands acreage affected by their projects, no matter how large or small, with equal, or often greater, numbers of preserved, restored, enhanced, or created wetlands acres. The mitigated wetlands generally have to be located on the same site, or at least in the same watershed, as the destroyed wetlands, and must consist of the same or similar wetland type.[6]

As a theoretical construct, the sequencing requirement is appealing: it steers development away from wetlands and then demands compensation for unavoidable impacts. But the reality is very different, due in large part to an additional regulatory provision cobbled onto the front-end of the permitting process.

Specifically, in its implementation of the Section 404 program, the Corps pairs the sequencing criteria with a requirement that applicants must demonstrate the absence of "practicable alternatives" to their proposed projects. That is, if any nonwetland alternative to an applicant's project—be it a different development configuration on the proposed site or development on an entirely different site—is deemed to be technologically, economically, and logistically feasible, would not affect the environment more adversely than the proposed project, and would achieve the project purpose, then the permit application will be denied. The mere existence of a practicable alternative, even if it is one involving property the applicant does not own, is sufficient to scotch any chance of obtaining a permit for the desired project.

The Corps, the resource agencies, and applicants spend staggering amounts of time—indeed, far too much time—analyzing the existence of practicable alternatives to projects at the expense of time that could be very well spent husbanding wetlands resources. By no means is this meant to suggest that regulation is an unnecessary frill. Rather, regulation establishes a baseline level of environmental protection that ensures no party will gain a competitive advantage by destroying the environment. But the Section 404 regulatory scheme is clearly, and imprudently, excessively front-loaded. The practicable alternatives process consumes the agencies and applicants resources but adds little by way of real environmental value; it is merely an expensive and time-consuming hurdle that drives many applicants from the process. Indeed, most individual applications end up being withdrawn for this or related reasons.[7] Those who stay the course *do* get permits eventually, subject to mitigation requirements. Given this fact (and given that driving people out of the process by overwhelming them with procedural requirements is bad government, not to mention unfair and ultimately counterproductive because it undermines the program), the choice becomes clear. We must streamline the Section 404 regulatory program by jettisoning the single-minded focus on practicable alternatives and refocusing our energies on the environmental side: namely, on minimization of impacts, mitigation of impacts, and enforcement of measures to compensate for impacts.

In short, while the Section 404 program's design conjures up, as typical end results of the mitigation process, visions of 50-acre freshwater marshes complete with nesting blue herons, common loons, and box turtles or vibrant 20-acre tidal estuaries teeming with shorebirds and native fish species, the current reality is much starker. In the field, these idyllic images are realized in few actual cases. Far more frequent are such familiar outcomes as the half acre of cattail marsh at the local shopping mall, the quarter acre of boggy soil surrounded by chain-link fence and full of soft drink cans near the local gas station, or the wet patch of marsh grass and wildflowers behind the local elementary school. Thanks to present mitigation policy, the country is strewn with these kinds of postage-stamp-sized, ecologically disjointed wetlands, whose capability to absorb flood waters, nurture endangered species, or perform other wetlands functions is minimal at best, and whose long-term sustainability is improbable.

The Section 404 program is supposed to prevent net losses of wetlands, but the tales of widespread mitigation failure suggest that the program is falling short of the mark.[8] Mitigation banking provides a creative way out of the Section 404 program's unimaginative and unsuccessful tit-for-tat mitigation cycle. Mitigation banks tend to cover relatively large areas, so they are more likely than tiny sites to retain their wetlands characteristics over the long term. They also tend to provide a wider range of functions and values than small wetlands and thus can be considered, as a general matter, to be of higher quality and greater value.[9] In short, mitigation banks are environmentally superior to many small, isolated wetland parcels. It would make more sense, therefore, for the government to encourage the channeling of mitigation monies into banking projects—that is, projects that work—than to continue on the present course of funding wetlands fragmentation and slow, but inevitable, destruction. Through mitigation banking, limited financial resources would go to preserve and restore the highest quality, most sustainable wetlands, to the great benefit of wetlands ecosystems across the nation.

Another creative solution involves efforts to control and eliminate nonnative plant species, which can ravage delicate wetland ecosystems. Such efforts are beginning to gain recognition as serving a valuable function. Precedent for this type of mitigation credit already exists in Florida. In Dade County, Florida, for example, applicants for certain types of permits are given an automatic option of contributing to the Dade County Melaleuca Eradication Fund and receiving mitigation credit.[10] Melaleuca is an aggressive water-loving plant that, because of its ability to outcompete native species and its rapid growth pattern, "is now regarded as the most serious threat to the integrity of all south Florida's natural systems."[11] Other invasive exotic species include Brazilian pepper, hydrilla, and water hyacinth. In California, EPA is considering awarding mitigation credits for the control of nonnative species of plants. (Stephanie Wilson, Wetlands Banking Coordinator, U.S. EPA, Region IX, Remarks at a Statewide Conference on Wetlands Mitigation Banking in California, July 29, 1994.)

Mitigation Banking Facilitates Enforcement

As the program is currently implemented, the Corps lacks resources to adequately monitor and enforce compliance at the many small, fragmented, on-site mitigation projects churned out under the existing regulatory scheme. Unfortunately, there is no agency follow-up for tens of thousands of permits that are issued each year. It is no secret, nor should it be surprising, that many compensatory mitigation failures are a direct result of poor enforcement. Plainly, enforcement is a vital part of any regulatory program, particularly here, where it is needed to protect the environment, and its absence is an invitation to abuse. Without enforcement, scofflaws have a field day.

Rigorous enforcement is also important for those who comply with the requirements—to demonstrate that their compliance does not disadvantage them in the marketplace. Ineffective enforcement undermines the credibility of the Section 404 program and does a serious disservice to developers who try in good faith to comply with the program's extensive requirements. The Corps's failure to identify and punish parties who flout federal authority acts as a negative incentive for other parties to conform to the law's parameters and, indeed, may place responsible parties at a competitive disadvantage.

Mitigation banking provides a solution to this problem. By collecting the Corps's follow-up obligations in fewer places, mitigation banking makes it easier for the Corps to monitor compliance and enforce violations. Government activity in this direction would do much to bolster the Section 404 program's credibility and prevent wetlands losses.

Mitigation Banking Provides Incentives to Conserve Wetlands: The Profit-Seeking Mitigation Banker

The Section 404 program is regularly criticized for its supposedly high mitigation failure rate. Undoubtedly, wetlands mitigation efforts, including mitigation banks, have sometimes failed to produce functioning, sustainable aquatic resources. These failures can be attributed to, among other things, hydrologic engineering difficulties, geologic obstructions or deficiencies in soil characteristics, consumption by wildlife, or simple neglect of the restored or created wetlands. Mitigation efforts are sometimes deemed failures because replacement wetlands are damaged by natural storm events. It is questionable, however, whether mitigation bank wetlands should be expected to withstand assaults from Mother Nature that natural wetlands themselves would not endure. In nature, wetlands are dynamic systems that move constantly, expanding and contracting according to hydrological, geological, and weather events, not to mention lawful off-site activities by humans as well as animals. Any program establishing mitigation banking should recognize this fact and take it into account in establishing the bank's long-term maintenance obligations.

In such cases, the nation experiences a net loss of wetlands. The high failure rates cited, however, are sometimes skewed by selective review of a few wetlands characteristics or are based on old data.[12] To the extent that failure does occur, our review suggests that it is more often caused by human shortcomings than by technical limitations. In particular, wetlands fail because (1) the applicant never begins or completes the required mitigation, (2) the Corps does not enforce compliance with the mitigation conditions, and/or (3) the mitigation is poorly designed or implemented.

Unquestionably, wetlands restoration and creation methodologies are complex and exacting; to be successful, mitigation projects must be comprehensively planned, scrupulously designed, rigorously implemented, and steadfastly maintained. This is a tall order for any nondivine entity to fill, let alone private developers engaged in the highly competitive business of constructing residential and commercial properties. Some argue, based on mitigation failure data, that all development in wetlands should be prohibited. That is unrealistic. As noted earlier, economic growth is inevitable and valuable, and the majority of applicants who persevere through the Section 404 process ultimately obtain permits.

Applicants whose projects are large enough to absorb the costs will hire expert consultants to design and carry out a high-quality mitigation plan. Such plans can and have been highly successful.[13] But most projects that affect wetlands are not sizable. A recent study of individual applications processed during 1992 found that half of all the applications involved 1.1 acres of wetlands impacts or less; 15 percent of the applications involved less than one-tenth of an acre.[14] Typically, these applicants do not themselves have the expertise to design and implement mitigation plans nor do they have the means to hire an expert consultant. The results are predictable.

So what to do? The answer is to seek better mitigation, not more mitigation or a blanket prohibition on all future wetlands alteration. To achieve better mitigation, the Section 404 program must be changed to create incentives for successful mitigation projects: incentives that will encourage private parties to restore, enhance, create, and preserve wetlands, as well as to experiment with and study wetlands restoration and creation techniques.

One "incentive," of course, is rigorous enforcement of mitigation obligations. As discussed earlier, the present front-loaded system, focused as it is on the alternatives analysis, gives enforcement short-shrift. Official recognition of mitigation banking will provide another powerful incentive. It will encourage the creation of entrepreneurial entities whose self-interest is intertwined with innovative mitigation techniques and successful, flourishing mitigation banks. Once people discover, with the government's blessing, that there is money to be made in the mitigation banking business, they will bring to bear all of their creative energies and will undoubtedly devise ways to improve mitigation methodologies.

For example, entrepreneurs would likely be inclined, at the outset of the program, to establish mitigation banks in areas where chances of success are high. It may be that not all wetland types are suitable candidates for mitigation using existing mitigation techniques. Degraded tidal wetlands are currently ideal candidates for mitigation banks, as wetlands hydrology—generally the most difficult wetlands component to engineer properly[15]—would already be largely in place. Pocosin wetlands and bottomland hardwood swamps may be less suited for successful mitigation using existing techniques, and so those wetlands types would be avoided, at least initially. Techniques for their successful creation and restoration would eventually be developed, in response to demand for wetlands credits possessing the functions and values of those wetland types.[16]

In most cases, the restoration, enhancement, or creation of existing or new wetlands is accepted as legitimate compensation for wetlands impacts. Compensation by an agreement for wetlands preservation, on the other hand, has met with a certain amount of resistance. Some argue that preservation does not fulfill the national goal of "no net loss" of wetlands and thus should not be permitted as compensatory mitigation for wetland impacts. The argument is that total wetlands acreage is reduced when development at one site is mitigated by preservation of existing wetlands at another site.

This reasoning assumes that a preservation agreement is never needed to preserve existing wetland acreage. To the contrary, however, without a preservation agreement some wetlands could be converted to upland through activities for which no permit is required. In those instances, there would be no mitigation requirement and no compensating creation of wetlands. For example, some wetlands may be threatened by off-site activities such as nearby development that lowers the water table, construction of levees that cut off sheet flow, or invasions of nonnative species. Other wetlands may be so degraded that they will, in the absence of human intervention, revert to upland. In each of these cases, preservation could be used, as a first step in a restorative process, to set aside and protect a site's wetlands functions. Acknowledgment of preservation's value as a tool in the Section 404 repertoire is beginning to creep into federal policy.[17]

Carefully guided, the invisible hand of the market can help achieve tangible increases in the quality and quantity of wetlands in the United States. With a market for the credits they produce, more people will take the entrepreneurial plunge into mitigation banking, thereby encouraging wetlands restoration, preservation, and creation and fulfilling a fundamental goal of the Section 404 program.

Under any mitigation banking model, some adjustment may be required to level the competitive playing field between public and private mitigation banks. In the wetlands context, government mitigation banks have an enormous competitive advantage. Public banks have not sunk thousands or millions of dollars into land acquisition ventures; their land already belongs to the public and thus is essentially "free." As a consequence, public banks can offer lower mitigation credit prices than private banks and still be profitable. But when the government enters the marketplace as a competitor, and not as a regulator, free market ground rules become skewed. To make the system work more fairly in such circumstances, the invisible hand must be tweaked somewhat to prevent economic chaos. Perhaps public bank mitigation credits should be assigned a handicap of sorts (similar to the handicaps carried by golf and polo players) to smooth out differences in price between public and private banks.

Mitigation Banking Provides Certainty to Developers and Regulators

Under the existing regulatory system, the government and developers are continually engaged in case-specific negotiations to establish and design mitigation conditions for individual projects. The negotiation and design processes are uncertain ones: it often takes weeks or even months longer than anticipated to negotiate mitigation conditions acceptable to all the agencies involved and the developer and to design wetlands systems that are suited to the particular conditions of an area. The negotiation and design processes are also costly ones, both in terms of direct outlays to hire hydrologists and biologists to design wetlands and indirect costs generated by unexpected delays in construction.

Mitigation banking eliminates the uncertainties and costs associated with the case-by-case negotiation and design of wetlands mitigation projects. Mitigation banking enables developer purchase of mitigation credits and subsequent scheduling of time- and capital-intensive construction activities without fear of expensive derailment by runaway permitting processes. Moreover, because the cost of mitigation credits at a bank would be established, developers' costs would not be subject to continuous run-up as a result of seemingly endless processing to determine the scope and composition of on-site mitigation requirements. By providing planning certainty to developers in these ways, mitigation banks can reduce the costs of the development process, and developers can pass those savings on to consumers. And by providing certainty to regulators that mitigation efforts will succeed, mitigation banks support the national goal of no net loss of wetlands.

The banking idea has been labeled by some as the "write a check and walk" option, implying that mitigation banks make it easy for developers to fill wetlands without concern for environmental consequences. However, this policy by no means allows developers to shirk their responsibilities to compensate for wetlands losses. The key question is not whether but how applicants should compensate for their impacts. Purchasing mitigation credits is not an empty exercise, particularly when the money goes to expert entities that are highly motivated to restore wetlands. The environment is better served if mitigation obligations are undertaken by experts—mitigation bankers—whose professional reputations depend on the mitigation outcome. And if the applicant's money goes directly to protecting the environment, rather than to the process, so much the better.

Mitigation Banking Provides Wetlands Conservation at No Public Cost, and Allows Private Landowners Reasonable Beneficial Use of Land

Great societal benefits can be reaped by motivating the private sector to provide products that serve the public interest. Through mitigation banking, the public's goal of conserving and enhancing wetlands resources can be attained at no cost to the public. Of course, the fact that what is being banked is "mitigation" necessarily contemplates that there will be wetlands loss. But if the loss has met the regulatory criteria and is therefore deemed to be in the public interest, on balance that loss has been determined to be justified by other gains in the public interest.

Mitigation banking safeguards the public fisc from the costs of acquiring large parcels of land and funding sophisticated wetlands restoration, creation, and maintenance activities there. In this way, mitigation banking provides for the conservation of wetlands that might not otherwise be protected at all due to budget constraints.

Moreover, mitigation banking may allow the landowner a reasonable beneficial use of a site that otherwise would be rendered useless by the Section 404 program's restrictions. That is, if a landowner applies for a permit to develop a site that includes wetlands, and the Corps denies the permit in order to protect the wetlands, the landowner may still realize some profit from the site if it can be used as a mitigation bank. This is an advantageous result because otherwise the landowner would be left with no use and the government would be vulnerable to a lawsuit to compensate the landowner for a regulatory taking.

In short, mitigation banking can effect a win-win: public benefit is obtained at virtually no public cost, and the landowner earns a reasonable return on his or her investment.

Conclusion

The first section of this chapter has addressed many of the attractive features of mitigation banking. Of course, banking is just one element in the Section 404 program; there are many other elements that must be addressed to make the program a fair and effective protector of wetlands. For example, while the present program is effective at steering people away from wetlands, it does nothing to encourage people to restore degraded wetlands. But encouraging such restoration should be a central component of any comprehensive wetlands preservation and conservation program. Mitigation banking is one among several ways to achieve that important goal. For the reasons explained in this section of this chapter, it should be encouraged.

An indirect benefit of having mitigation banking become a widespread reality would be the advancement of wetlands restoration and creation sciences. Advancements in these disciplines would auger well for humanity's ecosystem management capabilities, which are strained to the breaking point by our increasingly crowded world and thus are in great need of infusions of innovative ideas and technologies. Mitigation banking is just the mechanism to stimulate scientific progress, for the benefit of people and the environment alike.

An Environmentalist's Perspective—Time for a Reality Check

Jan Goldman-Carter and Grady McCallie

Introduction

Much has been made of the promise of mitigation banking. Its promoters reason that banking will provide more environmental benefits than past on-site mitigation practices because bank sites will be larger and more stable, they will be planned and implemented with more expertise, they will be constructed in advance of development activities impacting natural wetlands, and they will likely be easier to monitor and regulate. This reasoning is appealing superficially. However, it suggests only that banking may be an improvement over on-site wetland mitigation. On-site mitigation has proven a dismal failure and does not provide the standard by which mitigation banking should be judged. This reasoning also is purely speculative. The experience with mitigation banking is not adequate yet to confirm its promise. The promoters of mitigation banking have touted the promises of banking, but failed to face its pitfalls squarely. This section is intended as a "reality check" to highlight the potential problems that must be addressed before mitigation banking can be recognized as an acceptable means of replacing natural wetlands destroyed through development activity.

There is much at stake if mitigation banks become widespread substitutes for natural wetlands, and they fail. If this occurs, the result would be more natural wetlands lost, financial and administrative resources wasted, and the credibility of wetland regulatory programs eroded further. If mitigation banking is to facilitate, rather than undermine, the maintenance and restoration of the nation's wetlands, then regulators and lawmakers must evaluate mitigation banking objectively, with all its attendant risks as well as opportunities.

In defining the rules for mitigation banking, regulators and lawmakers also must be clear about their objectives. Ostensibly, the main objective is to provide the regulatory flexibility necessary to accommodate the use of mitigation banking where that approach provides the greatest ecological benefits. The underlying standard that must be satisfied is the full replacement of wetland functions and acreage lost due to development activity. However, much of the policy discussion suggests a different underlying objective: to promote private entrepreneurial mitigation banks to increase the supply of mitigation credits and thereby reduce the uncertainty and the cost of development activities in wetlands.

The rules of mitigation banking will differ considerably depending on which of these objectives are pursued. Mitigation banking is both risky and expensive. If mitigation banks are held accountable for full replacement of wetland functions and values and if mitigation credits reflect the full costs of wetland replacement, including the costs associated with the risk of failure and with perpetual care and protection, then mitigation banking is not likely to be an attractive business venture for many private bankers or an attractive mitigation option for many developers. Nevertheless, mitigation banking may produce a "win-win" solution for bankers, developers, and the public in certain limited situations. If mitigation banking rules are developed conservatively, consistent with the objective of maximizing ecological benefits and ensuring no net loss of wetland function and acreage, many environmentalists will be prepared to accept mitigation banking as a valuable wetland conservation tool.

On the other hand, mitigation banking rules that actively promote banking will almost certainly require the public to absorb many of the risks and costs associated with mitigation banking. Full replacement of lost wetland functions and values is likely to be sacrificed in many instances. Thus, in the interest of promoting mitigation banks, the public effectively will subsidize private development. Environmentalists will not support subsidizing—promoting—private destruction of wetlands for private financial gain. Mitigation should remain the financial responsibility of the developer benefiting from the destruction of public, natural wetland resources. To ensure that mitigation banking fully replaces lost wetland functions and values and that mitigation continues to be the financial and legal responsibility of the wetland developer, the potential pitfalls discussed herein must be addressed effectively.

Ultimately, the risks and costs of mitigation banking should limit effectively its application to those situations in which banking will (1) contribute to a broad-based ecosystem restoration project that has a high probability of producing significant net environmental benefits and (2) provide for some meaningful replacement of wetland functions and values lost due to the cumulative adverse effects of many small-scale wetland losses. Indeed, where mitigation banks are approved, compensatory mitigation should become mandatory for activities causing these small-scale wetland losses.

Specific Realities and Necessary Safeguards

Reality Number One: Wetlands Are Hard to Build and Failure Is Commonplace

"If resource agencies expect projects to replace all wetland functions exactly, the entire new generation of wetland restoration/creations will be judged failures," Paul Sugnet, Environmental Consultant.[18] At the very outset, mitigation banking arrangements must recognize the reality that virtually no replacement wetland will fully replace the natural wetland functions they are intended to offset.[19] Successful wetland mitigation, then, already assumes some retreat from a no-net-loss goal, at least unless wetlands are successfully replaced at more than a 1:1 ratio.[20]

Even when "success" is defined as something less than full replacement of lost wetland functions, mitigation failure is very commonplace. An estimated 50 percent of wetland restoration/creation projects have failed, based on short-term monitoring alone.[21] In most cases, the information is still not available to evaluate the long-term success of these projects.[22]

The reasons for mitigation failure typically arise from shortcomings in the art and science of restoration and mitigation, or from shortcomings in the institutional framework (e.g., adequate planning, monitoring, and enforcement of mitigation requirements), or some combination of the two.[23] Reality Number One is that the art and science of wetlands creation and restoration is in its infancy, and mitigation banking projects are as likely as not to fail for this reason alone.

To date, successful restoration has been documented for a few wetland types. Wetland creation has been judged successful in far fewer instances.[24] Tidal marshes and mangrove forests have been restored successfully along the Florida coast.[25] Similarly, estuarine marshes along the Pacific coast and vernal pools in California have been restored successfully.[26 27] There is some evidence of successful stream restoration in California, as well.[28]

Wetlands on agricultural land, including prairie potholes, can also be restored, often simply by breaking drainage tiles or plugging ditches to restore wetland hydrology.[29] Even in these relatively simple restorations, long-term cultivation often destroys wetland seed banks. As a result, some drier-end plant species vital to ground-nesting birds have proven difficult to bring back. Moreover, it is unclear whether prairie pothole restoration restores functions such as groundwater recharge.

The art and science of mitigation has been far less successful in restoring other wetland types. Northern peat bogs require centuries to mature and have proven virtually impossible to successfully restore, much less create.[30] Similarly, Carolina pocosin wetlands, also peatlands, have proven extremely difficult to restore. Their highly organic soils become oxidized when exposed to air, lowering the ground level. The altered elevation requires careful manipulation of water levels to establish wetland plant growth. Restoring wetland hydrology is further complicated by the already highly altered groundwater and surface water hydrology of pocosin wetlands.[31]

Bottomland hardwood wetlands have been successfully reforested, but not restored as wetlands. Flood control measures and other changes in the flow patterns of major rivers like the Mississippi have largely eliminated the natural hydrology of these forested wetlands, making full restoration nearly impossible. Even the more limited goal of reestablishing dominant tree species on the theory that the associated subdominant vegetation will follow has proven overly optimistic. At best, these reforestation projects may achieve a semblance of vegetative diversity in about 50 years.[32] Even then, the reforestation is not likely to produce wetlands.[33] Extensive changes in water regimes on western rivers similarly have destroyed the necessary hydrology for most riverine restoration.[34]

Even when mitigation projects attempt to restore those wetland systems most capable of restoration, they are by no means assured of success. Indeed, the potential for failure is high and the causes for failure are numerous. The most common failure is improper design or construction of the mitigation site's hydrology.[35] Establishing the proper wetland hydrology requires careful analysis of numerous hydrogeologic factors and careful supervision of site grading and water level management regimes.[36] A hydrologic system that requires frequent pumping or other manipulation of water levels is very likely to fail. One study of mitigation projects in south Florida found that 25 percent of the projects had "significant" hydrologic problems.[37] At least three mitigation banks have already failed due to faulty hydrologic design and construction.[38]

Another cause of failure is problems with compacted or clay soils, debris, fill, or contaminants at the mitigation site. At least two mitigation banks have already experienced performance problems due to substrate conditions that were not properly identified and addressed in bank planning.[39]

Establishing wetland vegetation has also proven difficult in application. Plant and/or seed selection, planting depths, and failure to properly sustain plants during the establishment phase have all been sources of mitigation project failure.[40]

These are simply the most common start-up problems in providing the basic ingredients of wetland hydrology, soils, and vegetation. Other common problems contributing to project failure include construction-related accidents (e.g., bulldozing or herbicide spraying the wetland seedlings), vandalism, oil spills, storms, invasion of exotic species, wildlife "eat outs" of newly planted vegetation, disease, and insect pests.[41]

Banking Safeguards that Recognize the Reality of Mitigation Bank Failure

Avoid Natural Wetland Losses. The proponents of mitigation banking can certainly dream of successful wetlands replacement, but they must plan for failure. Since full replacement of natural wetland function virtually is impossible and since many restoration and creation projects fail to even approximate lost wetland functions, the most essential safeguard is to prohibit the destruction of natural wetlands to the maximum extent practicable.

In the language of the Clean Water Act Section 404 program (and many state wetlands programs), wetland developers should be required first to avoid altering wetlands and to minimize wetland alterations that are required to achieve basic project purposes. Compensatory mitigation—whether on-site or off-site in the form of mitigation banking—should be employed only as a last-ditch effort to replace those wetland losses that are truly unavoidable.[42]

Limit Banking Projects to Restoration. Mitigation banking should be limited to restoration of former or degraded wetlands, where the basic ingredients for success—wetland hydrology, soils, and vegetation—are much more likely to be available. The chances for replacement of natural wetland function and value are improved substantially in the restoration context. Even in the restoration context, mitigation banks should not be approved for pocosin wetlands, bottomland hardwoods, and other wetland types until successful restoration of the these wetland types has been proven. Regulators must resist the temptation to lower the success criteria hurdles in order to ensure that wetland consultants and credit producers can clear them. The standard for success must continue to be full functional replacement.

Require Mature Bank Credits in Advance of Development Activities in Natural Wetlands. Mitigation bank credits should not be sold or traded (thereby allowing for natural wetland alterations to proceed) until the replacement wetland conditions on which the credits are based reach full maturity. In other words, mitigation credits must be fully established in advance of the sanctioning of natural wetlands losses. Advance mitigation is the simplest and safest way to insure against bank failure.

Require Approval of Detailed Site Plans in Advance of Bank Approval. Mitigation banks must not be approved without advance review of detailed, site-specific mitigation plans by regulators, resource agencies, wetlands experts, and the public. As noted earlier, mitigation failure is often caused by lack of adequate information and improper design and planning. Approved plans, along with performance and design standards and success criteria must be incorporated into the banking agreement and wetland permit.

Require Careful Monitoring of Bank Operations. Mitigation bank construction and management must also be subject to careful monitoring by bank operators and regulators, resource agencies, and the public. Funds and resources must be available for independent monitoring by regulators. The public must be assured access to monitoring information.

Require Approved Contingency Plans and Financial Assurances. Mitigation banks must not be approved without adequate contingency plans for reestablishing bank credits upon failure and financial assurances that will act both as an incentive for full performance and a source of funding upon bank failure. These contingency plans and financial assurances must be incorporated into the banking agreement and authorizing wetland permit.

Regulators must demonstrate a clear commitment and capability to enforce bank compliance. Regulators must not approve a mitigation bank unless they are confident that they can and will enforce full performance of the bank to achieve full replacement of lost wetland functions and acreage. The public must also have the capacity to enforce banking agreement violations (see Reality Number Four discussion later).

Reality Number Two: Restored and Created Wetlands Are Hard to Maintain Long-Term, and Proof of Long-Term Success Is Nonexistent

Even when mitigation projects achieve short-term success in establishing hydrology and vegetation, long-term success is by no means assured. Restored wetland ecosystems, even those that appear healthy and successful after the initial one—three growing seasons, can suffer collapse five, eight, even ten years later. Fires, floods, storms, invasion of exotic or aggressive species, and consumption of vegetation by wildlife all can jeopardize the long-term replacement of wetland functions and values. Natural wetlands, like other ecosystems, have many feedback loops that are not even recognized, much less well understood. Consequently, correcting failures in managed ecosystems involves a great deal of unknowns.[43] In the words of wetland scientist William Odum, "It is clear that in most cases, we stand little chance of predicting the long-term development of restored and created wetlands until we better understand the process in natural wetlands. Only in the intertidal coastal wetlands do we appear to have assumed any measure of control and predictability."[44]

Cattail intrusion is one well-documented example of problems that can cause long-term failure in created and restored systems. If not checked, cattails can outcompete other wetland vegetation and create a monoculture with significantly reduced habitat values. While cattail intrusion can occur in natural wetlands, it appears to be particularly common in created wetlands.[45]

Storm events are also very likely to contribute to project failure. As the Environmental Law Institute (ELI) study so aptly notes, "We can expect the occurrence of a ten-year storm event, even a 100-year storm event, during the life of a bank. If the bank has not been designed for these events, it is operating in a fantasyland."[46]

Beyond these natural occurrences, adjacent land uses can also jeopardize the long-term maintenance of wetlands. Adjacent farming, logging, or construction, for example, can cause rapid siltation of bank wetlands. Siltation, in turn, can affect a range of wetland functions significantly, including flood storage, pollution reduction, plant species diversity and fish and wildlife habitat. Polluted run-off from farm fields, highways, and industrial sites can similarly destroy wetland functions. Uses permitted at the banking site following initial establishment, such as logging, grazing, haying, or intensive recreational activities (e.g., off-road vehicle use), can diminish or destroy wetland functions in much the same way.

One of the most disturbing aspects of moving forward quickly with mitigation banking is how little is known about the long-term performance of restored and created wetlands. California Coastal Conservancy attorney Reed Holderman noted at a mitigation banking conference in 1994 that "Only after twenty years of data on the same site will we really know if created wetlands can replace ancient ones."[47] Mitigation banking must progress slowly and with adequate safeguards to ensure against a mitigation banking crisis of at least the same dimensions as the savings and loan crisis of the 1980s and early 1990s. Only this time, there can be no federal bailout.

Additional Safeguards Recognizing the Long-Term Uncertainty of Mitigation Banks: Hedging Our Bets

The safeguards against start-up and near-term failure of mitigation banks discussed earlier also will help to forestall long-term bank failure. However, the gist of Reality Number Two is that much of nature, as well as human activity, is beyond the bank operator's knowledge and control.

So, in addition to the prudent measures already described, the logical response to such uncertainty is to "hedge our bets."

One of the major conclusions of ELI's mitigation banking study is the need for a transition to mitigation banking through the use of pilot banking projects. For the three major types of mitigation banks, public works banks, public general use banks, and private entrepreneurial banks, the ELI study concludes that:

> Current experience has revealed a great deal about public works banks—not all of it affirming the approaches that have been used. . . . *There is virtually no useful experience with public general use and private entrepreneurial banks; thus, the focus of the pilot programs should be on these banks.* The transition to broader use of banking need not be lengthy, but it should be pursued intentionally so that valuable lessons in banking structure can be applied rather than relying entirely upon ad hoc decisions in the field. *The ad hoc approach is partly what has brought us to the current situation in which there is little banking, and great uncertainty among regulators, the public, and prospective credit producers alike.* (emphasis added)[48]

Mitigation banking projects should be recognized as experimental and should be approved and implemented on a pilot project basis. Pilot projects should focus on wetland restoration projects that have a high likelihood of producing clearly demonstrable net increases in wetland function and acreage. Likely examples would seem to be banks that restore tidal and estuarine emergent marshes and mangroves as part of broad publicly supported plans to restore critical, but highly degraded, ecosystems (e.g., San Francisco Bay Area, the Everglades and Florida Bay, and the New Jersey Hackensack Meadowlands).

Restoration of isolated agricultural wetlands would also seem a likely prospect. Such projects seem to have a relatively high success rate at relatively low cost and can produce significant water quality, flood storage, and wildlife habitat benefits. The lessons learned from these pilot projects can then be applied to similar mitigation banking projects that follow. While this approach will not remove the difficulties and uncertainty of long-term wetland replacement, it should help reduce both the economic and ecological risks of these endeavors.

Reality Number Three: Measuring the Loss and Replacement of Wetland Function is Difficult, and Existing Methodologies Are Not Reliable

Reality Number Three is a sobering one: how do we establish a trade in wetland debits and credits when we have no uniform currency? Asked another way, how do we compare the "apples" of lost wetland acreage and function with the "oranges" of replacement wetlands? How can we possibly ensure "no net loss" if we cannot establish a reliable method for equating the two? The sobering reality is that mitigation banking cannot achieve no net loss without such a methodology—and it is currently not available.

Quantifying the functions of a natural wetland slated for the bulldozer is a complex proposition. Wetland functions will differ depending on wetland type and extent, level of degradation, adjacent land uses, and location within the watershed. The measurement of various wetland functions, and the relative values attributed to these functions, will differ according to the methodologies, assumptions, and subjective value judgments employed by the analyst. These wetland functions and acres to be lost are quantified, albeit somewhat subjectively and imprecisely, as wetland debits.

It is the next stage of the analysis that becomes particularly precarious: measuring the replacement wetland credits and equating the debits from one wetland system with the credits of another. Here again, the replacement wetland functions will differ depending on wetland type and extent, level of degradation, adjacent land uses, and location within the watershed. In addition, the ecological functions already existing at the replacement site, wetland or otherwise, must not be measured as credits, since they were preexisting at the site and were not produced by the bank. Indeed, to the extent that these ecological functions are eliminated in order to produce credits, the credits should be discounted. Credits to be produced by the bank must also be realistically discounted to account for the risk of failure inherent in the banking venture. Even if this analysis can be accomplished with any degree of consistency and credibility, the equation is still not complete. We now have established a number of wetland debits and credits, but we have not yet established a credible conversion between the two. We are still comparing apples and oranges.

These seemingly insurmountable problems of quantifying and equating debits and credits can be reduced by eliminating some of the variables outlined earlier. Obviously, the same assessment methodology and assumptions can and must be used to measure debits and credits. "In-kind" wetland replacement (i.e., the replacement wetlands are the same type as the debit wetland) certainly improves the chances of a meaningful comparison of debits and credits. Measuring fully mature credits (i.e., the function and value of replacement wetlands that have been successfully established in advance) reduces the need to discount credits due to risk of failure (though longer-term failure must still be accounted for). Nevertheless, the variables are still many and confounding.

Many evaluation methodologies have been, and are continuing to be, developed. These include very simple indices of value, such as acreage or species diversity, as well as much more complex methodologies for measuring various ecological and hydrological wetland functions.[49] While much attention is being paid to this quandary, the fact remains that it has not been solved. As the ELI study concludes, "[I]n order for a wetland mitigation bank credit currency to work, it must be (1) simple to determine and to monitor and (2) able to represent a sufficient range of values and functions. None of the existing systems do both of these things well. . . . Finally, adjustments in the currency may be necessary over time if the expected results of compensation are not in fact being produced by the system. Banking instruments will need to provide for this potentiality; most currently do not."[50]

Additional Safeguards Recognizing the Reality of Crude Comparisons

The reality is that we are not likely ever to establish a neat, precise, and reliable system for comparing quantitatively the natural wetland functions lost to development with the net wetland functions replaced through mitigation banks. Therefore, if mitigation banking is to proceed with a genuine goal of no net loss, then the uncertainties in the equation must be reduced where possible and, when in doubt, wetland replacement must err on the side of producing excess wetland functions, rather than a wetland function deficit.

The uncertainties can be reduced in several ways, several of which have already been discussed. First, the emerging assessment methodologies can be tested and refined through close monitoring of pilot banking projects. Second, banking instruments should be drafted to provide for credit adjustments when the predicted results fall short of actual compensation requirements. Uncertainty can also be reduced by selling credits only when fully established and mature and by generally replacing wetlands in-kind and within the same watershed. In addition, uncertainty is reduced by approving banks that restore those wetland types that have a relatively high restoration success rate and restore specific wetland systems that will likely produce very significant net ecological benefits.

Compensation ratios greater that 1:1 are frequently used, and must be used, to help ensure full replacement of wetland functions despite all of the uncertainties surrounding mitigation banking. Many wetland mitigation banking schemes and policy documents already employ this approach. For example, the Florida Department of Environmental Regulation (DER) has established a "sliding scale" of ratios for several mitigation banks, taking into account degradation of the impacted wetland, form of compensation, and stage of success of the mitigation wetland. A 1:1 compensation ratio may be employed where a severely degraded mangrove wetland is replaced by a fully successful replacement mangrove wetland. At the other end of the scale, a saltmarsh enhancement project may have to replace impacted natural saltmarsh at a ratio of 10:1.

To account for uncertainties of mitigation banking, compensation ratios, even for most in-kind restoration projects, must exceed 1:1 and should exceed 2:1 in most cases. More speculative projects, such as creation projects, out-of-kind projects, and projects selling credits prior to full maturity, should trade credits at higher ratios. Finally, enhancement and preservation projects that contribute less significant net ecological benefits should trade credits at even higher compensation.[51]

Reality Number Four: The Legal Mechanisms Are Not in Place to Ensure the Success and Perpetual Protection of Mitigation Bank Wetlands

Specific banking site plans, performance and design standards, monitoring requirements, and contingency plans all help reduce the risk of mitigation failure—if they are enforceable. However, disturbing questions remain about enforceability of bank enabling instruments. Who is liable for bank failure? The permittee who has purchased wetland credits as a mitigation bank client? The credit producer? The bank operator? The long-term owner of the banking site? What performance or damages are any of these parties liable for? Who can enforce compliance with the bank enabling instrument? Is that instrument enforceable, either legally or practically? As critical as these questions are, most of them have not been convincingly resolved in the establishment of existing banks or in the proposal of new ones.

The most logical party to assume liability for bank failure is the credit producer who undertook to provide the wetland credits, brought together the resources to do so, and should have planned for contingencies in the event of bank failure.[52] However, other options could include the permittee/client, the bank operator or long-term site owner, the regulatory agency, or no one at all. Liability for bank failure should be clearly assigned, yet many existing bank instruments are silent on liability.[53]

Assuming liability is clearly assigned, probably to the credit producer (who is often the client, as well), the next critical question is what exactly are they liable for? Are they strictly liable for achieving short-term design and/or performance standards? If these are achieved, and the bank fails five or ten years later due to unanticipated errors in the original design, is the credit producer liable for correcting the problem and rehabilitating the site? Who is responsible for bank failure once the credits are sold and the site is turned over to a long-term site owner? The limits of liability must be defined clearly to protect both the bank and the public. Private mitigation banks will certainly seek to limit their liability in order to reduce their risk and exposure to high costs of rehabilitating bank sites that fall short of design and performance standards.[54] Yet, limiting liability to short-term compliance with design or performance standards or success criteria leaves no one liable in the event of longer-term mitigation failure. The public is likely to be left "holding the bag."

The next logical question is who will enforce compliance by the credit producer? The Corps of Engineers is the logical "enforcer" of mitigation obligations stemming from its enforcement of Section 404 permits. However, it is uncertain whether the Corps could enforce the typical enabling instrument: a Memorandum of Agreement (MOA). As a practical matter, the uncertainty on this issue alone likely would forestall a successful Corps enforcement action based on an MOA. To help ensure enforceability then, the 404 permit for the bank should include the detailed conditions and protections of the enabling instrument.

Even if the Corps has the legal authority to enforce the banking instrument, questions remain about the agency's commitment and resources for doing so, as well as the agency's ability to force bank compliance as a practical matter. The Corps's track record for enforcing 404 permit conditions is generally poor, in large part due to limited staff resources and an emphasis on permit issuance rather than permit enforcement.[55]

Beyond the issue of resources and commitment, there are practical questions of forcing actual performance by the bank. This is true particularly where the development activity has already been completed and the bank credits sold or traded. In such cases, neither the bank client (the permittee) nor the credit producer has an incentive to comply with the bank requirements.

Despite the poor record of on-site mitigation, the high risk of mitigation failure, and the myriad of uncertainties that surround enforcement of banking agreements, many existing bank instruments do not even address enforcement. Some have been in debit status for some time with no corrective action and no enforcement.[56] Several government-operated (Department of Transportation—DOT) banks have been among those that failed to comply with provisions of the banking agreements.

Reality Number Four is a stark one: already banks are failing due to lack of enforcement, yet banks continue to be approved without effective enforcement mechanisms in place.

Legal Safeguards to Ensure Bank Performance and Perpetual Protection

Mitigation banks must not be approved until the issues of liability, enforceability of the enabling instrument, and remedies upon bank failure are resolved clearly. Bank approval should involve a 404 permit to the credit producer and bank operator for the establishment and operation of the bank itself. The permit should embody or incorporate by reference detailed banking provisions, including a detailed site plan, performance and design standards, success criteria, contingency plans, monitoring requirements, financial assurances, and specific arrangements for liquidating financial assurances, levying penalties for noncompliance, and trust funds for long-term site maintenance.

Financial assurances must be required that (1) ensure that funds will be available to repair and maintain the bank site in the event that problems are not corrected by the credit producer and (2) provide an incentive to design, construct, and maintain the bank site to achieve full wetland replacement in perpetuity. DOT and other government-operated banks also must provide such financial assurances, possibly in the form of a legislatively enacted dedicated fund. The Corps must be able to draw on these financial assurances readily upon its determination that a violation has occurred and not been corrected. Despite the availability and flexibility of financial assurances as an enforcement tool, few existing or proposed banks require them.[57] This failure to protect the public interest amounts to a breach of agency public trust responsibilities. This must change.

Enforcement of banking violations should be strengthened by establishing clearly a citizen suit remedy to enforce banking agreements. The Clean Water Act (CWA) provides for citizen suits to enforce Corps and EPA actions related to 404 permitting. It is unclear, however, whether and to what extent the CWA's existing citizen suit provision would provide a remedy to enforce mitigation banking agreements.

In the bank permit, deed restrictions should be placed on the banking site to preclude, in perpetuity, any inconsistent use and provide for effective enforcement of those deed restrictions. Moreover, the permit must require establishment of a trust or escrow fund that will finance the long-term management of the site, ensuring the continued functioning of the replacement wetlands.

Reality Number Five: Mitigation Banks Are Difficult to Bankroll without Speculating in Credit Futures

There is an inherent "Catch-22" in financing mitigation banks, particularly for privately funded banks. Financing is required up-front to establish the bank, yet the revenue stream from the bank comes only when the wetland credits produced are sold. For privately funded banks, the only way to finance the bank may be to sell credits before they mature. Yet, given the high risk of bank failure discussed earlier, this practice amounts to speculating in credit futures.

According to authors Shabman, Scodari, and King, premature wetland credit sales must be allowed for private, market-based mitigation banks to operate: "This concern [risk of mitigation failure] may tempt regulators to require private commercial bank mitigations to be in place and fully functioning before they could be used as compensatory mitigation. Use of this risk-minimizing strategy in the credit market context would force private banks to bear the full costs of waiting for the maturation of replacement wetlands (i.e., opportunity costs of invested capital) as well as all failure-risk costs. However, these costs would probably be too high for most private commercial banks to earn a competitive return on investment. If a market-based trading system is to operate, there must be opportunities for private banks to sell credits before replacement wetlands reach functional maturity or self-maintenance, and in some cases, perhaps even at the time mitigation is initiated." [58]

Consequently, it seems that mitigation banking policymakers must choose between promoting private entrepreneurial mitigation banks by allowing premature credit sales (and reducing the bank's exposure to risk, as well) and ensuring against the risk of mitigation bank failure by requiring the full maturity of bank credits prior to sale and prior to development activity in wetlands.

Nonprofit banks must labor under the same financing Catch-22 as private entrepreneurial banks, except that they are not driven by the search for a "competitive rate of return." Instead, they must be assured of a sufficient rate of return to adequately cover all their costs, including long-term landholding and maintenance costs and insuring against the risks of failure. Often, nonprofit banks are financed by a number of previously identified clients, such as development companies or local government authorities. Few nonprofit organizations are able to provide the up-front costs of bank development themselves.

Government banks are usually established to mitigate for public works projects; highway department banks to mitigate for highway project wetland impacts are the primary example. These DOT banks are typically funded through state and federal highway administration funds. State

DOTs can justify carrying the costs of one or more mitigation banks because they can easily anticipate that the bank credits will be needed to offset future losses due to highway construction.[59]

Government banks could also be established to mitigate for private development activities. In such cases, the sale of wetland credits should be priced ultimately to recoup the costs of establishing and maintaining the mitigation bank, including the "opportunity costs of invested capital" born by the government during the maturation of the banking credits. Government banks would nevertheless require substantial up-front public funding through taxation, permit fees, bond issues, or other similar measures.

The problems of up-front financing have also led mitigation bank promoters to look to public lands and public resource agencies to provide banking sites and technical expertise in wetland restoration and creation. In many cases, these public lands and resources have been dedicated for the very purpose of wetland restoration, enhancement, or preservation over and above that provided by mitigation banking. The use of these resources to establish mitigation banks is an outright government subsidy to wetland developers—a subsidy opposed by environmentalists.

The up-front financing of mitigation banks becomes particularly problematic when the start-up costs are high, the wetland maturation period is long, and/or the risks of demand for credit and of bank failure are particularly high. Forested wetland restoration and creation is an example of a wetland type that is impacted severely by development and has very important wetland functions. Restoration will, in many cases, require costly efforts to reestablish wetland hydrology. A diverse, mature, fully functioning forested wetland system will require some 50 years of maintenance before wetland credits would mature fully.

With forested wetland restoration, Reality Number Five comes into sharp focus. It is extremely unlikely that a private entrepreneurial or nonprofit bank could achieve the requisite return on investment from establishing a forested wetland restoration site if the wetland credits from the site cannot be sold or traded until 50 years hence. The uncertainty surrounding mitigation requirements (and therefore the demand for these mitigation credits) makes it even more difficult to imagine such an investment. Ultimately, it seems, the public will be forced to bear the risks associated with any forested wetland banking projects. It will be forced, at a minimum, to bear the risk that the demand for credits 50 years hence does not recoup the cost of the initial public investment. More likely, the public will bear the risks and costs of bank failure as well.

Safeguards that Limit Speculation and Reduce the Risk of Bank Failure

The most prudent approach to reducing the risks of bank failure is to require full maturation of bank wetlands in advance of development that impacts wetlands. Credits should be priced to reflect the true costs of full wetland replacement, including the opportunity costs of the invested capital pending full maturation. Until vast improvements are made in the track record for bank success and in effective enforcement of banking agreements, the prudent approach is the only one that will avoid widespread bank failure. The same prudent approach is, of course, warranted for on-site mitigation.

In addition, problems in financing mitigation banks must not be overcome through public subsidies. Public lands and public resources otherwise committed to wetland restoration must not be used to subsidize mitigation banks.

Undoubtedly, it is true that the prudent approach will dampen the interest of entrepreneurs in the fledgling mitigation banking industry. It may well be that mitigation banks for forested wetland systems may have to be government subsidized if they are to be constructed at all. If this is the case, it is all the more reason why these wetland systems should not be destroyed in the first place and why extraordinary efforts should be taken to avoid impacts to these precious and virtually irreplaceable resources.

Summary and Conclusions

Mitigation banking does indeed hold promise as an improvement over on-site mitigation efforts to date. However, the performance of mitigation banks must be judged by the unrelenting standard of full wetland replacement, not the dismally low standard of on-site mitigation to date. Consequently, the first step toward promoting mitigation banking must be a fresh commitment to applying consistently the Section 404(b)(1) guidelines to both on-site and off-site mitigation. In other words, all the same safeguards identified earlier for mitigation banking must be required now for on-site mitigation to "level the playing field" between these two forms of mitigation.

The mitigation banking debate must be an honest one. We must face squarely the reality that successful mitigation is generally expensive and the risk of failure is high. Rather than lower these hurdles to promote the use of private banks, we must recognize the need to move gradually and to recognize the limitations of mitigation banking in the short term. Unless we impose the safeguards identified herein to reduce the chances of mitigation failure, the mitigation banking "industry" will surely make a farce of federal and state wetlands protection programs.

Yet, it seems unlikely that bankers who faithfully apply these safeguards will find banking to be a lucrative business venture. It seems equally unlikely that most land developers who faithfully abide by these safeguards will find mitigation banking a cost-effective option for them. Given these constraints, environmentally responsible mitigation banking seems to hold promise in only limited circumstances.

First, mitigation banks are not likely to function successfully as for-profit business ventures. A private nonprofit or a public bank operator is much more likely to succeed without a profit motive.

In addition, mitigation banking seems to hold promise as part of a larger ecosystem restoration effort, widely recognized as producing large-scale environmental benefits (e.g., restoration of Everglades/Florida Bay hydrology, ecology, and water quality). In this scenario, wetland losses are authorized and mitigated only in a manner consistent with a comprehensive sustainable development plan.

Mitigation banking also seems to hold promise when established to offset the cumulative impacts of numerous very small, unavoidable impacts in the watershed—unavoidable impacts that have thus far gone unmitigated because of the lack of feasible mitigation options. Indeed, wherever such a bank exists, compensatory mitigation should be mandatory for all activities that are authorized under general permits. No longer can compensation for such activities be credibly viewed as infeasible.

How do environmentalists feel about mitigation banking? Some see the potential for genuinely successful wetland restoration. But environmentalists are also skeptical. Environmentalists have seen the sham of on-site mitigation projects that are never even started, much less successfully completed. If mitigation banking is developed conservatively, with the appropriate safeguards and as a means of adding demonstrable net ecological value, many environmentalists might well be prepared to become believers.

Notes

1. Federal Guidance for the Establishment, Use, and Operation of Mitigation Banks, 60 Fed. Reg. 12,286 (March 6, 1995).

2. See, e.g., Environmental Law Institute, *Wetland Mitigation Banking* (23): 117–23 (1993) [hereinafter *ELI Report*]; Leonard Shabman et al., U.S.

Army Corps of Engineers, IWR Report No. 94-WMB-3, *Expanding Opportunities for Successful Wetland Mitigation: The Private Credit Market Alternative,* pp. 15–21 (Jan. 1994).

3. Ibid. John Meagher and Marjorie Wesley, "Encouraging Stewardship on Private Lands," *National Wetlands Newsletter*, Mar./Apr. 1994, at 13; see S. 851, 104th Cong., 1st Sess. § 2(a)(4) (1995). Moreover, a significant percentage of endangered or threatened species of plants and animals and other wildlife are directly or indirectly dependent on wetlands for survival. See *Hearings on H.R. 961 Before the Subcomm. on Water Resources and Environment of the House Comm. on Transportation and Infrastructure,* 104th Cong., 1st Sess. (1995) [hereinafter *Hearings on H.R. 961*] (statement of Jan Goldman-Carter, National Wildlife Federation) (citing National Wildlife Federation, *Endangered Species, Endangered Wetlands: Life on the Edge* 7 (1992)).

4. *Hearings on H.R. 961, supra* note 3 (statement of Robert Perciasepe, Assistant Administrator for Water, U.S. EPA).

5. See U.S. EPA and Department of the Army, *Memorandum of Agreement Concerning the Determination of Mitigation Under the Clean Water Act Section 404(b)(1) Guidelines* 3–4 (Feb. 6, 1990) [hereinafter *Mitigation MOA*].

6. See 60 Fed. Reg. at 12,291; see also U.S. EPA and Department of the Army, *Establishment and Use of Wetlands Mitigation Banks in the Clean Water Act Section 404 Regulatory Program* 147 (Aug. 23, 1993) (interim guidance); Mitigation MOA, *supra* note 5, p. 4.

7. See, e.g., Virginia S. Albrecht and Bernard N. Goode, *Wetland Regulation in the Real World* 26 (Feb. 1994).

8. Cf. Shabman et al., *supra* note 2, at 1 ("[T]he record of success for on-site mitigation is spotty, and there is widespread concern that net losses of jurisdictional wetlands are continuing.") (citations omitted.)

9. See, e.g., Cathleen Short, U.S. Fish and Wildlife Serv., Biological Report No. 88(41), *Mitigation Banking,* pp. 2–3 (July 1988).

10. See Environmental Law Inst. & Inst. for Water Resources, U.S. Army Corps of Engineers, IWR Report No. 94-WMB-2, *Wetland Mitigation Banking: Resource Document,* pp. 97–98 (Jan. 1994); Telephone Interview with Kathy Fanning, Wetlands Planning Program, Dade County, Florida (Oct. 14, 1994).

11. Office of Technology Assessment, *Harmful Non-Indigenous Species in the United States* 261 (Sept. 1993).

12. See, e.g., Charlene D'Avanzo, "Long-Term Evaluation of Wetland Creation Projects," in *Wetland Creation and Restoration: The Status of the Science,* Jon A. Kusler and Mary E. Kentula, eds., pp. 76–82 (U.S. EPA Doc. No. EPA/600/3-89/038b, Oct. 1989).

13. See, e.g., Roy R. Lewis III, "Creation and Restoration of Coastal Plain Wetlands in Florida," in *Wetland Creation and Restoration: The Status of the Science,* Jon A. Kusler and Mary E. Kentula, eds., pp. 73–91 (U.S. EPA Doc. No. EPA/600/3-89/038a, Oct. 1989).

14. Albrecht and Goode, *supra* note 7, pp. 20, 22.

15. See *ELI Report, supra* note 2, pp. 95–96.

16. For example, some entrepreneurs and the government are already experimenting with techniques for restoring bottomland hardwood swamps and pocosin wetlands. In North Carolina, Triangle Wetland Consultants is engaged in an effort to restore 25 acres of bald cypress and tupelo gum swamp, and the U.S. Fish and Wildlife Service is working to restore thousands of acres of pocosins at the Pocosin Lakes National Wildlife Refuge. See Lawrence S. Earley, *Can We Build a Wetland?,* Wildlife in North Carolina, July 1994, pp. 9–13.

17. See, e.g., 60 Fed. Reg. pp. 12,288-89; *Mitigation MOA, supra* note 5, at 4.

18. Sugnet, Paul, "How Successful Have We Been in Restoring and Creating Wetlands?" Course Materials for the Ninth Annual Conference on Wetlands Law and Regulation, May 1994 (Sugnet 1994).

19. "National Wetland Mitigation Study: Wetland Mitigation Banking," IWR Report 94-WMB-6, prepared by the Environmental Law Institute, February 1994, p. 27 (*ELI Report*), citing *Wetland Creation and Restoration: The Status of the Science, Volumes I & II,* Jon A. Kusler and M.E. Kentula, eds., Washington, DC: Island Press, 1990; and King, Dennis M., "Wetland Creation and Restoration: An Integrated Framework for Evaluating Costs, Expected Results and Compensation Ratios," Prepared by Chesapeake Biological Laboratory, University of Maryland, Solomon, Maryland, for U.S. Environmental Protection Agency, Office of Planning and Evaluation (1991).

20. *ELI Report* , p. 76.

21. Sugnet, Paul, "How Successful Have We Been in Restoring and Creating Wetlands?" Course Materials for the Ninth Annual Conference on Wetlands Law and Regulation, May 1994 (Sugnet 1994) citing Conservation Foundation, "Protecting America's Wetlands: An Action Agenda," 69 pp. (1988).

22. Id. See also, Roy Lewis, "Creation and Restoration of Coastal Plain Wetlands in Florida," in Kusler and Kentula, pp. 73–101.

23. See, *ELI Report*, chapter 8; Kusler and Kentula; National Research Council Report, "Restoration of Aquatic Ecosystems: Science, Technology and Public Policy," National Academy of Science, Washington, DC, (1992).

24. Id. See also, Redmond, Ann, "Report on Mitigation in Florida State Permitting Efforts," 1990, Florida Department of Environmental Regulation, Tallahassee, FL cited in *ELI Report.*

25. Lewis, *supra* at note 22.

26. Josselyn, M., J. Zedler, and T. Griswold, "Wetland Mitigation Along the Pacific Coast of the United States," in Kusler and Kentula, *supra*.

27. Sugnet, *supra* citing Coe, T. U.S. Army Corps of Engineers, letter dated March 18, 1994 (determination that 69 constructed vernal pools met permit success criteria); Personal communication with Mark Littlefield, U.S. Fish and Wildlife Service, Sacramento Field Office.

28. Clay, D. H., "High mountain meadow restoration," in *California Riparian Systems: Ecology, Conservation, and Productive Management,* R. E. Warner and K. M. Hendrix, eds., pp. 477–79 (Calif. Water Resources Report 55, 1984).

29. See, e.g., Madsen, C., "Wetland Restoration in Western Minnesota," in *Increasing Our Wetland Resources,* compiled and edited by John Zelazny and Scott Feierabend, p. 93 (National Wildlife Federation, 1988).

30. *ELI Report.* p. 28.

31. Personal communication, Russ Lea, wetland scientist.

32. Haynes, R., and L. Moore, "Reestablishment of Bottomland Hardwoods within National Wildlife Refuges in the Southeast," in *Increasing Our Wetland Resources,* at 101 (1988); Sharitz, R. "Bottomland Hardwood Wetland Restoration in the Mississippi Drainage," in *Restoration of Aquatic Ecosystems,* p. 486; Personal communication with Jim Allen, U.S. Fish and Wildlife Service Southern Science Center.

33. Personal Communication with Greg Obol, National Biological Survey, Fort Collins, CO; and with Russ Lea.

34. See, e.g., Cooper, D. J., "Mountain Wetland Vegetation Dynamics," in K. M. Mutz, et al. (technical coordinators) *Restoration, Creation, and Management of Wetland and Riparian Ecosystems in the American West,* 23 (1988); S. W. Carothers and G. S. Mills, "The creation and restoration of riparian habitat in southwestern arid and semi-arid regions," in Kusler and Kentula, *supra,* pp. 351–366.

35. See, *ELI Report,* pp. 58 and 77.

36. Id.; Hollinds, G. G., G. E. Hollis, and J. S. Larson, "Science base for freshwater wetlands mitigation in the glaciated northeast United States: hydrology," in *Mitigating Freshwater Wetland Alterations in the Glaciated Northeastern United States: An Assessment of the Science Base,* J.S. Larson and C. Neill, eds., pp 131–43, (1986).

37. *ELI Report* at 77, citing Erwin, K., "An Evaluation of Wetland Mitigation within the South Florida Water Management District," vol. 1, Kevin L. Erwin Consulting Ecologist, Inc., Fort Myers, FL, 1991.

38. *ELI Report,* pp. 77–78.

39. Id.

40. Id.

41. Id.

42. See 40 CFR Part 230; Corps and EPA Mitigation MOA (1990).

43. Ehrenfeld, D. 1992. "Managed Systems," in *Ecology, Ethics and Environment: The Broken Circle,* Stephen Kellert and Herbert Bormann (eds.), Yale University Press, cited in *ELI Report,* p. 26.

44. Odum, W. "Predicting Ecosystem Development Following Creation and Restoration of Wetlands," in *Increasing Our Wetland Resources,* p. 67 (1987).

45. Personal Communication with Joy Zedler; Brown, Stephen C., "Avifaunal Use of Restored Wetland Habitats," *Great Lakes Wetlands* 5(4) (Winter 1994); Lewis, *supra* note 22, pp. 97–98; Morrison, D. et al., "Effects of a Freshwater Discharge from Finger Canals on Estuarine Seagrass and Mangrove Ecosystems in Southwest Florida" in *Proceedings of the 17th Annual Conference on Wetlands Restoration and Creation,* p. 115 (1990).

46. *ELI Report,* p. 81.

47. Holderman, R., "Case Studies in California: Lessons Learned," outline for presentation, Wetlands Banking Conference, University of California, Davis, CA. July 29, 1994; see also, Lewis, *supra* note 22.

48. *ELI Report,* p. 128.

49. *ELI Report,* ch. 7.

50. Id., pp. 74–75.

51. Id., p. 76, citing Castelle, A. J. et al., "Wetlands Mitigation Banking," cited in report prepared for the Shorelands and Coastal Zone Management Program, Washington State Department of Ecology, Olympia, WA. Publication No. 92–12, 37 pp. and appendices (March 1992).

52. *ELI Report,* pp. 83–84.

53. Id.

54. See, e.g., Shabman, et al., *supra* note 2 at Executive Summary xi; Garbisch, in Kusler and Kentula, cited in *ELI Report,* p. 80.

55. See, e.g., U.S. General Accounting Office, "Wetlands Protection: The Scope of the Section 404 Program Remains Uncertain," GAO/RCED-93-26. April 6, 1993, p. 25.

56. *ELI Report,* p. 88.

57. *ELI Report,* p. 85.

58. Shabman et al *supra* note 2 at Executive Summary, x.

59. *ELI Report,* p. 100.

6

Wetland Mitigation Banking Markets

Leonard A. Shabman, Paul Scodari,
and Dennis King

Private credit markets result from a special type of mitigation banking that likely will dominate future banking activity.* The vast majority of mitigation banks in operation today are single-user banks; that is, each was developed by a single large public or private developer to provide only for its own future mitigation needs. By contrast, private credit markets would develop if entrepreneurs create credits for sale to developers in need of compensatory mitigation. Such private commercial banks could help the nation achieve no net loss of wetlands by increasing the opportunity to obtain successful compensatory mitigation for permitted wetland losses.

Federal wetland regulations state that the use of off-site mitigation banks may be an acceptable alternative in certain situations. Yet, rela-

*This chapter was adapted from a report prepared for the U.S. Army Corps of Engineers (Corps), Institute for Water Resources, as part of the Corps's National Wetland Mitigation Banking Study. The chapter findings and recommendations are the authors' alone and do not represent the position of the Department of the Army.

tively few mitigation banks have been established thus far, despite their potential to increase the efficiency and effectiveness of wetland regulation. This is because traditional single-user banking arrangements necessarily are limited to those large public and private developers that routinely undertake many independent or linear development projects and can afford a substantial up-front investment in compensatory mitigation.

Mitigation credit markets offer the opportunity to increase the efficiency and effectiveness of compensatory mitigation by providing the banking option to a wider set of permit applicants. Indeed, toward this end a number of states and localities across the nation have established public commercial banks and public fee-based mitigation systems (sometimes referred to as "in-lieu" fee systems). Public commercial banks offer mitigation credits for sale to the general public and use the proceeds from credit sales to recoup the costs of bank construction and management. Similarly, public fee-based systems charge permit fees for projects involving small wetland impacts in lieu of the direct provision of mitigation by permittees. Fee revenues are accumulated in trust funds for the intended future provision of replacement wetlands by the government entity. While the broader establishment of these two types of public mitigation systems could potentially extend the advantages of mitigation banking to a wider set of permit applicants, important obstacles must first be overcome. One major problem for establishing public banks involves the substantial up-front public financing needed for bank construction and management. Public fee-based systems may also face financing problems since there is no guarantee that fee revenues accumulated in trust funds for replacement wetlands will not be diverted to other uses. Moreover, both types of public mitigation systems face the risk that fee (credit) charges will be insufficient to cover the full costs of providing compensatory mitigation for the fill activity they serve.

Unlike commercial mitigation banking by public entities, a private credit market system would tap the profit motive to encourage private entrepreneurs to produce mitigation credits with private capital. If entrepreneurs emerge to sell credits to many possible buyers, a private market for wetland functions would develop. Market competition could ensure that mitigation credits were provided at least cost, as well as provide incentives for the further development of wetland restoration technologies as credit supply firms seek out more successful mitigation techniques.

The most obvious benefit from private credit market systems is the opportunity to secure mitigation for the many small wetland impacts that would otherwise go unmitigated. For example, under general permits, including nationwide 26 permits, compensatory mitigation is often not required when wetland alterations are so small that the possibility of on-site mitigation is deemed impractical or infeasible. The cumulative impact of many such small wetland losses is one cause of slippage from the no-net-loss goal. The widespread establishment of private credit market systems could correct this deficiency by making credits available for sale in small increments. Regulators could then require compensatory mitigation in cases involving small wetland impacts by having developers purchase equivalent credits from established private commercial banks.

Credit market systems could also have broader application to permitted development projects involving more significant wetland impacts. Current wetland regulations emphasize the on-site mitigation option in the hope that important site-specific wetland functions, such as stormwater retention and erosion control, will be retained at the site affected by the fill activity. However, wetland development projects also impact wildlife habitat and ecological "life-support" functions that may be transferable to other locations within watersheds.

In fact, the opportunity to replace lost habitat and life-support functions successfully may often be improved by conducting mitigation away from the development site. For example, in some instances the inflexibility in the mitigation sequencing rules of the regulatory program—which require permit applicants to avoid, minimize, and mitigate wetland impacts on-site—may limit the possibility of successful mitigation as well as wetland preservation. This can occur if permitting decisions pay too little attention to the possible fragmentation, isolation, and functional degradation of replacement wetlands provided by in-kind and on-site mitigation, or wetlands preserved as a result of avoidance and impact minimization. The important implication is that when the wetland functions lost as a result of permitted development are largely transferable within the watershed, it may be desirable to secure compensatory mitigation through private commercial banks.

Allowing the purchase of private market credits in certain cases as a substitute for on-site mitigation could also enable regulators to avoid the several sources of failure associated with on-site mitigation efforts. Foremost among these are problems of enforcement:

1. When permits are granted conditional on the provision of mitigation, typically "on-site and in-kind," often no compensation effort is ever made.
2. If mitigation is initiated, regulators often do not have the technical expertise nor the time to check the mitigation plans for technical quality and feasibility or to check the construction practices that execute plans.
3. Often there are too few resources to allow for regulatory monitoring of mitigation projects that are constructed.
4. If a mitigation project is monitored and determined to have failed, there may be no responsible party liable for rectifying that failure.

5. If a mitigation project is constructed and judged successful in the short term, often there is no assurance that the mitigation site will be maintained as a wetland into the future.

The credit market alternative could greatly reduce the institutional and ecological sources of mitigation failure inherent in the current regulatory program by leading to the following outcomes.

1. Private credit markets would tap and combine mitigation expertise, planning, and capital in a manner that is typically not possible with on-site mitigation projects. Then, if a permit applicant had the option of buying credits from an established bank that had already planned for and provided replacement wetlands, there would be less chance that the permit applicant's compensatory mitigation requirement would go unfulfilled.

2. The consolidated mitigation projects provided by private banks would enable the regulatory agency to concentrate its limited oversight and monitoring resources on a much smaller number of mitigation sites.

3. Regulators would have more leverage and a greater variety of tools for imposing cost liability for mitigation failure in the banking option since regulators could dictate the conditions under which banks could create and sell credits.

4. Private banks would reduce the problem of ecologically vulnerable mitigation sites by consolidating what would otherwise be many isolated and fragmented on-site mitigation projects into a relatively few areas of replacement wetlands that could be sited and constructed according to watershed goals.

5. The reality of successful replacement wetlands and available mitigation credits would make the negotiation of permit applications more focused on issues concerning the need for the permit and the ecological value of the impacted wetland if the permit is or is not granted. These important permitting issues would then be divorced from concerns about the possibility and likelihood of successful mitigation.

Indeed, these advantages have been recognized by entrepreneurs and wetland regulators in many areas of the country, and as of the start of 1993, two private commercial mitigation banks had already obtained federal permission to create and sell mitigation credits under the Section 404 regulatory program. Moreover, across the nation the challenge of creating regulations conducive to private credit market systems is actively being discussed in a number of states and localities.

Conditions Necessary for the Widespread Emergence and Ecologic Success of Private Credit Markets

Of course, there are localities and circumstances where credit markets cannot improve prospects for successful mitigation. Where suitable restoration sites or sources of water for wetland restoration projects are not available, for example, producing mitigation credits may be impossible. Where wetland development is not profitable enough for permit seekers to afford high-quality mitigation the demand for credits may be too small for the credit market alternative to succeed. However, prospects for successful mitigation credit markets are limited in most cases by the same geophysical and economic conditions that limit opportunities for successful mitigation of any kind. In general the opportunities for mitigation credit markets to help further the no-net-loss goal exist wherever mitigation is viewed as an acceptable alternative to prohibiting all wetland development.

A ready supply of mitigation credits would emerge from entrepreneurs in many areas of the country provided that the conditions for market operation established by regulators enabled credit suppliers to earn a competitive return on investment. Wetland regulators, however, have legitimate concerns about whether the bank mitigation projects from which credits are sold will succeed over time. The emergence of the private market alternative and its ability to improve the effectiveness of compensatory mitigation depends on the capacity of regulators to fashion trading and regulatory rules that provide enforceable environmental safeguards without being cost-prohibitive.

Wetland Regulation and Mitigation Credit Markets

The operation of private credit markets to improve opportunities for achieving successful wetlands mitigation requires the interaction of three agents: credit suppliers, permit applicants, and regulators. Each has its own objectives and constraints, and each approaches mitigation credit trading with its own expectations and strategies. To a large extent the opportunities and constraints faced by credit suppliers and permit applicants depend on regulatory goals and the trading rules established by regulators to achieve them. The ability of mitigation credit markets to meet the objectives of all three groups will determine whether or not they can operate to provide compensatory mitigation.

The objective of permit applicants is to maximize the rate of return on investments in wetland development projects. To the extent that permit applicants are required by regulators to provide mitigation, they will try to minimize the costs of this requirement so as to maximize development returns. The objective of credit suppliers is to maximize the rate of return on investments in wetland restoration. They will try to minimize their costs of producing mitigation credits so as to maximize their own return on investment.

If a permit applicant buys credits from a supplier to meet its mitigation requirement, what the applicant is really purchasing is not mitigation, per se, but a development permit. That is, the willingness of permit applicants to pay for credits is established by the regulatory requirement for compensatory mitigation as a condition for receiving permits. Because of this, a normal market exchange between a permit applicant and a credit supplier cannot be expected to result in assured, long-term mitigation success. In the absence of any conditions imposed by regulators to minimize the risk of mitigation failure, there is no economic incentive in this exchange for permit applicants to strive for self-maintaining wetlands as a mitigation product. In fact, since the potential profits of permit applicants is inversely related to mitigation costs, there is an economic incentive to minimize mitigation costs and, therefore, mitigation quality.

The poor success rate observed for on-site mitigation efforts reflects in large part these poor incentives for successful wetland restorations (see National Research Council, 1992). The existing market for on-site mitigation illustrates that when regulators do not establish adequate design standards, enforce actual construction, or hold permit applicants (or mitigation suppliers) liable for mitigation project failure, permit applicants can and often will reduce restoration expenditures at the expense of long-term mitigation success.

The objective of regulators is to protect the wetland functions in a watershed. The Clean Water Act's Section 404 regulatory program has administratively adopted a policy goal of achieving no net loss in wetland function, to be followed by net gain, to meet this objective. These goals are the result of legal mandates that govern the administration of the regulatory program.

The different objectives of credit suppliers, permit applicants, and regulators offer the potential for deal-making, which is the essence of markets. The challenge facing regulators is to establish a framework for the operation of credit market systems in which the goals of all three agents can be met: credit suppliers and permit applicants each earn some profit, and the no net loss goal of regulators is achieved. The very existence and structure of markets in wetland mitigation credits depend on regulatory policies.

The supply and demand conditions in markets for mitigation credits are exceptional because of the two roles that must be played by government. First, credit markets could not exist in the absence of government regulations that create the demand for wetland development permits and make the granting of permits conditional on compensatory mitigation. Second, with regard to requirements for compensatory mitigation, permit applicants are price-conscious but not quality-conscious; their only concern is whether mitigation satisfies permit conditions established by regulators. It is the regulator, not the buyer of mitigation, who must impose "quality control" on the market through trading rules establishing how and when credits can be created and sold (King, 1992).

Figure 6.1 illustrates the various ways in which regulatory policies influence the underlying forces of supply and demand in private credit markets. The left-hand column identifies the factors underlying the supply of mitigation credits, and the right-hand column identifies the factors underlying the demand for credits. The supply of credits reflects the costs of acquiring (or leasing) and restoring former wetland areas to provide mitigation. The demand for credits is derived from the demand for permits and reflects the value of credits to permit applicants.

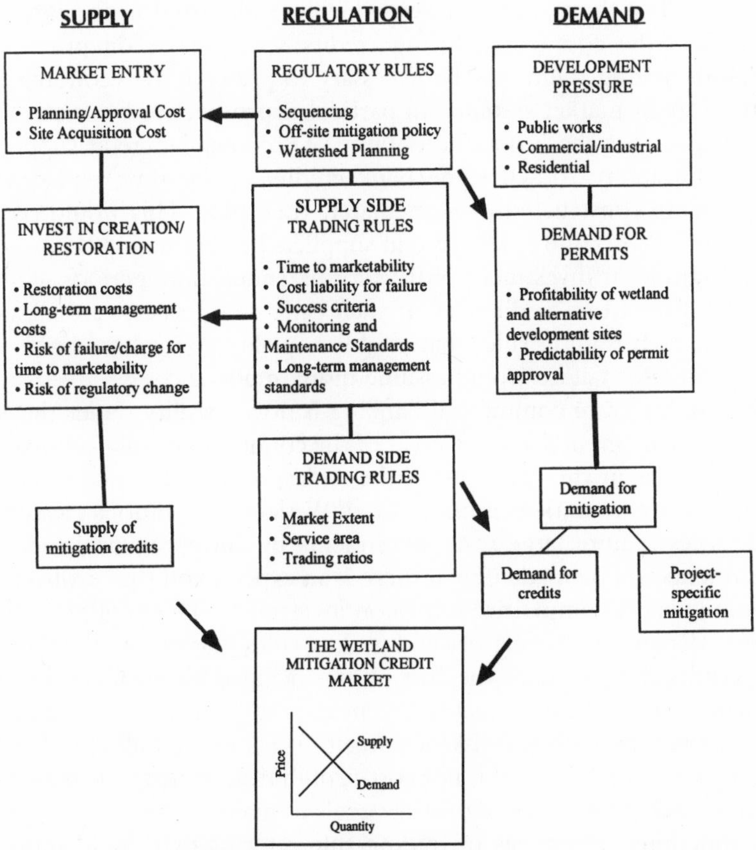

Figure 6.1 How Regulatory Policies Affect Prospects for the Emergence and Success of Mitigation Credit Markets

The center column of figure 6.1 identifies policy decisions that influence the underlying forces affecting either the supply of or demand for credits. Lines connecting the policy column with the supply and demand columns indicate where regulatory policies have the most significant impact. The policy column includes government decisions regarding regulatory rules and trading rules.

The regulatory rules include policy decisions regarding entry into the credit supply business and watershed planning. The effects of regulatory policies on credit demand and supply are explored in a later section.

Regulator concerns with credit trading center around the risk of mitigation failure. To address these concerns in the establishment and use of private commercial banks, regulators can establish a set of interrelated *trading rules* to increase the probability of mitigation success and thus the certainty with which policy goals can be met through credit market systems. All of these trading rules, which are discussed in the next section, affect the cost of producing credits, and thus credit prices, and trade-offs among them may be necessary to preserve the economic viability of credit market systems. In particular, prospective bank entrepreneurs argue that the economic viability of commercial credit supply requires the ability to sell some bank credits at the time replacement wetlands are constructed or immediately thereafter. This would reduce the financial costs and risks to credit suppliers by eliminating the need to tie up significant investment outlays for extended time periods without any cash flow from credit sales. If such early credit sales were allowed, however, then trading rules that establish quality control standards and cost liability for failure would assume more importance.

Although private commercial banking is now a reality under the Section 404 program in few small areas of the country and will likely expand to other areas in the near future, the widespread emergence of private mitigation credit markets is not assured. Wetland restorations created for credit sales require large-scale investments by entrepreneurs, and such investments will be made only if there is an expectation that profits from sales will yield a competitive return on investment. This profit potential, in turn, depends on regulatory and trading rules that dictate the demand for permits and influence the cost of producing mitigation credits. The challenge confronting regulatory agencies is to set rules for credit trading systems that limit the risk of mitigation failure and allocate liability for failure in a manner that is not cost-prohibitive, while at the same time to ensure achievement of regulatory goals to maintain and improve wetland functions. The types of trading-rule reforms that could promote this result are the subject of the next section.

Trading Rules to Promote the Economic Viability and Environmental Success of Private Credit Markets

Concerns about market-based trading systems center around the potential for restoration failure. Figure 6.2, which shows the "restoration success" time-path for a mitigation site, illustrates the nature of this concern. The vertical axis of figure 6.2 defines the functional value per acre at a mitigation site, and the horizontal axis measures time, where t=n is the time at which the mitigation site is constructed, t=n+1 is the time at which the site reaches a self-maintaining state, and t=n+2 is the time at which the site achieves functional maturity. When the site reaches a self-maintaining state, full function and value have not (necessarily) been achieved, but the site has a high degree of persistence and resilience to natural and anthropogenic disturbances and does not require extensive management inputs to stay viable. The solid line shows the time-path representing how the level of ecological functioning of a restoration site increases with time. The dashed lines represent confidence bands around the time-path. The confidence bands narrow over time as restoration "success" becomes more certain. In terms of the regulator's concerns, the confidence bands show that the probability of restoration failure declines with time (King, Bohlen, and Adler, 1993).

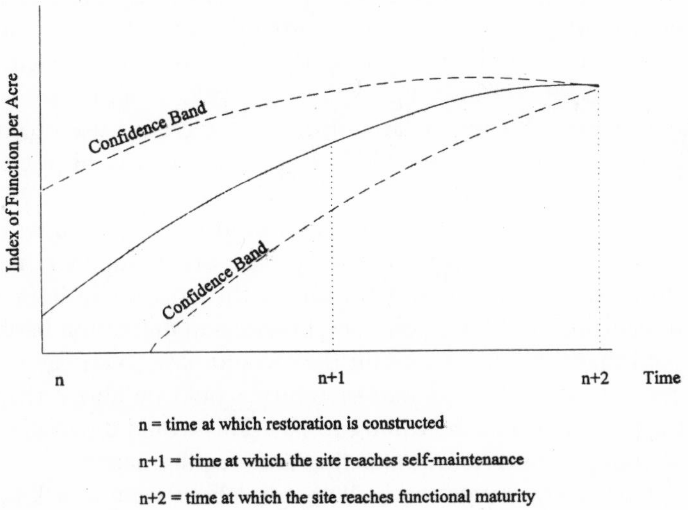

Figure 6.2 Change in Confidence in Restoration Success Over Time

Concerns over project failure, and who is liable for such failure, are heightened by the disappointing historical record of on-site mitigation efforts. However, many who are skeptical about wetland restoration in the mitigation context fail to distinguish between failures of the science and failures due to poor application of the science. The available evidence suggests that much of the observed failure of on-site mitigation is the result of vague restoration goals, inadequate expertise in performing restorations, failure to fully implement and enforce mitigation requirements, and an absence of site monitoring and management over time. This suggests that the institutional problems that lead to mitigation failure and net loss of wetlands should be addressed in setting up a market-based mitigation trading system but should not be confused with the technical challenges of wetland restoration.

Timing of Credit Marketability

In general, the Corps and EPA prefer that bank replacement wetlands be in place and functioning before credits can be used to offset permitted wetland impacts. This timing requirement stems from regulator concerns about mitigation project failure. At the same time, however, the timing issue is critical for the economic viability of private commercial banks. If regulators prohibit credit sales until fully functioning or self-maintaining wetlands have been achieved at mitigation sites, then credit suppliers would bear two costs: (1) costs of waiting for the maturation of replacement wetlands (i.e., the opportunity costs of invested capital) and, (2) costs of self-insuring against the risk of unforeseen natural events that might disrupt the attainment of the criteria used to measure success (assuming they are stated by the regulators). Credit prices would need to rise to cover all these costs.

In most cases, the cost of waiting and bearing strict liability for mitigation failure would be too high for private credit suppliers to earn a competitive return on investment. Given the potentially long waiting times to gain approval for credit sales, investors in mitigation banking are concerned that the price per credit they would have to charge to ensure a competitive, risk-adjusted rate of return would be above that which permit applicants would be willing to pay. This would especially be true if on-site mitigation does not face the same requirements.

The March 1995 mitigation banking guidelines seem to acknowledge this problem. The guidelines state that:

> Since financial considerations are particularly critical in early stages of bank development, it may be appropriate to allow limited debiting

> based upon a projected level of aquatic functions at a bank (e.g., 15 percent of the total credits projected for the bank at maturity).[1]

This provision provides one possible way in which regulators' concerns for mitigation project failure can be reconciled with the financial constraints of private credit suppliers.

However, even this added flexibility in the timing of credit sales may be insufficient to promote the widespread establishment of private commercial banks. Some entrepreneurs would not enter the credit supply business unless they were permitted to produce credits concurrently with the wetland impacts for which the credits will serve as compensatory mitigation. In fact, the federal permit for the Florida Wetlandsbank allows for such concurrent mitigation. In this case the banker successfully argued to regulators that the financial viability of the venture depended on using revenues from credit sales to finance the construction of replacement wetlands for those credits. The bank does intend to provide some advanced mitigation once sufficient revenues from credit sales based on concurrent mitigations have been accumulated.

At the Millhaven Plantation bank, credits can be released incrementally as mitigation activity is completed for a particular mitigation parcel. The bank will be allowed to sell one-half of the total mitigation credits generated by that parcel following a "preliminary determination of hydrology" that is conducted according to federally approved specifications. The bank must then show within three years that the parcel satisfies wetland delineation criteria relating to hydrology, soils, and vegetation before the remaining credits generated by the parcel can be sold. Moreover, the Corps project manager for the bank has sole discretion to establish trading ratios for any particular use of the bank. The Corps will adjust trading ratios to account for the maturity of replacement wetlands relative to impacted wetlands.

Regulatory and resource agency field staff consented to the bank provision for concurrent mitigation based on their recognition of the substantial costs of restoring the site up-front and the need for the bank to proceed with site restoration in a phased manner. The Corps was confident that the banker's favorable restoration site, plan, and mitigation expertise would result in a more successful, and more easily monitored and maintained, mitigation than what is typically provided by on-site mitigation projects.

The Millhaven Plantation bank and the Florida Wetlandsbank illustrate that if market-based trading systems are to operate, there may need to be opportunities for credit suppliers to sell credits before full functional maturity or self-maintenance is reached at wetland restoration sites and, in some cases, perhaps at the time in which mitigation is undertaken or even earlier. However, such early credit sales will be allowed only if regulators' concerns about the risk of project failure and who bears the consequences of failure are accounted for.

The consequence of mitigation failure is either that compensation for granting a permit is not realized or that the public must make an expenditure to repair the failed mitigation. In advance of the replacement wetland being in place and fully functional, the failure risk cost for a mitigation site is the product of (1) the probability that the restoration site will not achieve some long-term functional maturity and (2) the cost to repair or replace the restoration site when the compensation is not achieved or does not persist over time.

Performance standards must be established to judge whether a bank mitigation is failing or has failed, and these should be defined in advance of credit sales. In addition, performance standards should provide some leeway to account for less-than-extreme natural events that may cause a bank mitigation to evolve along a somewhat different path than originally planned.

The success criteria written into the permits for the Millhaven Plantation bank and the Florida Wetlandsbank appear to provide some flexibility by focusing on biological diversity at mitigation sites. Millhaven Plantation is required to maintain 300 trees per acre, and at least 25 percent of the "dominant" trees must be hardwoods. Further, no single species of planted or naturally occurring tree can at any time represent more than 30 percent of the dominant trees. Similarly, the success criteria for Florida Wetlandsbank require 85 percent survivorship of planted vegetation at two and five years after a mitigation parcel is certified for credit sales. Florida Wetlandsbank's permit goes on to say that "No more than 10 percent of the planted area may support exotic or undesirable plant species; it is noted that 10 percent of exotic or undesirable plant species may contribute to habitat diversity."

Each of these two permitted commercial banks are held to their respective performance standards during the course of five-year monitoring and maintenance periods established for each mitigation parcel certified for credit sales. Each bank is required to perform site monitoring and submit monitoring data to regulators as well as remedial plans for any discovered deficiency. In the case of Millhaven Plantation, if a deficiency is uncovered, a new five-year monitoring and maintenance period begins at the completion of remedial work undertaken to correct the deficiency. Both banks are released from further responsibility for any mitigation parcel in which the five-year monitoring and maintenance period is successfully completed.

As the permits for these two banks implicitly acknowledge, it would be unreasonable to hold credit suppliers to performance standards for more than some limited period of time. One of the concerns, however, is that after all credits are sold and performance periods are successfully completed, there will be no interest in maintaining the mitigation site. This concern has two elements. One is that the site will require long-term management, the second is that the owners of the site will seek to convert it to another use at some future date.

The contract provisions that authorize mitigation suppliers to create and sell credits could address these potential problems. For example, contracts might require that restoration projects be designed to be self-maintaining and/or there may be a requirement for some form of endowment with the earnings dedicated to perpetual maintenance. The endowment might be put in the hands of a management agency or a conservation group that would have similar maintenance responsibilities as a parks department. The ability to sell the site for a nonwetlands use might be restricted by requiring either a plan to transfer the site to public ownership and/or a conservation management entity, through permanent easements and deed restrictions.

The permits for Millhaven Plantation and Florida Wetlandsbank include such provisions. The land on which the Millhaven Planation bank is located is owned by a private, second party who leases the site to the banker. The permit for the bank is conditional on a perpetual conservation easement with the Corps, which requires the landowner to observe certain management standards designed to ensure the future status of the mitigation site as a wetland area. The Florida Wetlandsbank also leases the bank site from a separate landowner—the city of Pembroke Pines. Mitigation areas for this bank are protected by conservation easements into perpetuity, which also require the city to perform perpetual site management. Payments of $1,000 per mitigation acre were provided by the banker, based on estimates of maintenance cost jointly agreed to by the banker and regulator.

Regulators must clarify the "contract" conditions for credit suppliers in Memoranda of Agreement (MOA) and/or bank permits. The agreements recorded in these contracts must specify (in addition to mitigation siting, design, and construction plans): performance standards that define the conditions under which mitigation projects would be judged successful; monitoring and maintenance requirements to uncover and correct deficiencies, and provisions for long-term site management. Moreover, performance standards should provide some leeway to account for less-than-extreme natural events that might cause mitigation sites to evolve along somewhat different paths than originally planned.

Permit applicants (credit demanders) are unconcerned with the failure of mitigation banks, unless they are held liable for any costs necessary to repair a failed restoration. And while many credit suppliers likely would try to ensure that their restoration sites are successful in order to further their future prospects in the credit supply business, the risk of failure may be a concern to more opportunistic credit suppliers only if restoration projects fail before all credits are sold or if they are liable for mitigation project failure. This suggests that to ensure quality control at mitigation banks, regulators should impose cost liability on credit suppliers for failure to meet site design, performance, and management standards.

However, cost liability should not be imposed for mitigation failures resulting from natural disasters or other extreme events that prevent the attainment of performance standards for completed mitigation parcels. If credit suppliers were held liable for mitigation failures resulting from extreme events beyond their control, this could raise the risk costs borne by credit suppliers to the point where credit market systems could not operate. To ensure mitigation quality control while maintaining the economic viability of private credit markets, regulators should allocate to credit suppliers those failure risk costs resulting from nonperformance with contract requirements regarding the design, performance, and management of mitigation projects but not those costs resulting from extreme events.

This issue was explicitly recognized and accounted for in the Florida Wetlandsbank. The bank's permit specifies that if "acts of war, acts of God, rebellion, strikes, or natural disaster, including hurricane, flood, or fire" prevent the attainment of bank performance standards, the banker will not be held liable for such mitigation failure. However, the permit also says that if such extreme events "do not preclude the bank from performing permit conditions, the bank shall not be relieved of its obligations under the permit." While the permit for Millhaven Plantation has no similar provision, the Corps will not require the bank to replant vegetation destroyed in any completed mitigation phase as a result of extreme natural events, such as hurricane damage, but the bank would be required to fix any damaged water-control structure.

The degree of liability imposed for failure of a mitigation bank should reflect realistic failure probabilities and repair costs. If this occurs, the private entrepreneurs' profit motive will encourage them to use current restoration technologies carefully and encourage them to develop new technologies in order to reduce the burden of liability. Factors to be considered in estimating the probability of failure and the repair cost for any mitigation site should include:

1. The stringency of requirements established by regulators for restoration design, performance, and management at the mitigation site. The more stringent the requirements, the lower the probability of failure and the less the cost to repair a failed site.
2. The qualifications of, and regulators' historical experience with, the restoration contractor at the mitigation site. The more skilled and experienced the restoration contractor, the lower the failure probability.
3. The point in the time-path from initial restoration construction to functional maturity at which the credit sale is made. As time passes the certainty of successful restoration increases and costs to repair a failure falls.
4. The location of the restoration site within the larger watershed system. Placement of the site in the watershed where hydrology and potential biological integration is greatest suggests a higher probability of success.
5. The particular wetland type being restored at the mitigation site and historical restoration success rates associated with this wetland type. These factors can be used to judge likely restoration success.
6. The security of the long-term status of the site as a wetland. Easements and trust funds for perpetual management increase probability of success over the long term.

There are at least four options available to regulators for allocating cost liability for controllable failure risks to credit suppliers or demanders. Such liability mechanisms, which are described herein, should be included in the contracts that regulators write for each bank or wetland development permit. The regulator should choose among the options (not use all of them) in recognition of the expected failure probability at the site. Further, as is illustrated in the examples provided by the two permitted private commercial banks, liability mechanisms must be adjusted to the estimated failure probabilities and expected repair costs for each situation.

Higher Trading Ratio The regulatory agency may adjust the trading ratio for credits from the bank parcel to address controllable failure risks. The trading ratio required for any particular sale to a permit applicant would be based on some computation of the likelihood of restoration project failure. For example, assuming that the regulator seeks to achieve a no-net-loss goal, a trading ratio for failure risk from purchases at one site might be 2:1. All other factors equal, that ratio would imply a failure probability of 50 percent at the credit supplier's site and also that such failure would be complete (i.e., no functional value increase would occur at the site). Different trading ratios may be required for different mitigation sites or parcels to account for different failure probabilities across sites or parcels. In a competitive market, private credit suppliers would want regulators to impose lower trading ratios for any particular trade, and to this end would seek to reduce failure risk. And the lower the trading ratio required, other factors equal, the lower the compensation cost that would be paid by the permit applicant.

This option imposes risk costs on credit purchasers (i.e., permit applicants), but once the trading ratio is set and the credits are purchased, the public sector would be accepting the risk cost of restoration failure. Higher trading ratios would raise the costs to permit applicants of securing permits, and may dampen the demand for permits, and then for credits, to the point where the credit market would not operate. Therefore, the ratios must be based on realistic failure probabilities and repair costs (see the six items listed earlier).

Performance Bonds The regulatory agency may alternatively require credit suppliers to post performance bonds as a way to provide financial assurance. With this option, the bond requirement would be set by and paid to the regulatory agency, and the payment would be reimbursed with interest if at some future date the regulator certifies that the credit supplier's mitigation was successful. Partial refunds would be available for partial restoration success at the credit supplier's site. In fact, there may be some way to justify partial refunds each year. Early credit sales from the site would be permitted, and if the site fails the money in the bond would be used by the regulator to repair the mitigation project. The amount of the bond for any particular case would reflect the regulator's best estimate of the cost to repair the mitigation site if it fails; failure probability is not the concern of the regulator in setting the bond amount.

This approach places the failure risk cost on the credit supplier who would be expected to pass this cost on to customers. The total risk cost borne by a credit supplier is the sum of two costs. One cost is the difference between the market interest rate the supplier pays on the bond amount and the amount of interest (if any) paid by the regulator holding the bond. If the two interest rates are equal, this cost is zero. The second cost is measured by the credit supplier's expected probability of nonreimbursement times the amount of the bond. For the credit supplier, the expected probability of nonreimbursement should be possible to assess if the contract with the regulatory agency clearly specifies the conditions under which site failure would be established. Well-specified criteria for defining mitigation failure would increase the credit supplier's ability to estimate and take actions to minimize failure probability.

Performance bonding is the financial assurance approach used in both the Millhaven Plantation bank and the Florida Wetlandsbank. The permit for Millhaven Plantation requires that the bank post a $5,000 bond with the Corps for each acre of mitigation for which a "preliminary determination of hydrology" is made. Once the Corps makes a final determination that these acres have been restored to their "predrained hydrology," the bond amounts will then be reduced to $1,000, and a five-year monitoring and maintenance period begins. The bond balance will then be released to the banker upon completion of the monitoring and maintenance period only if no negative reports regarding the restored acres are filed by the relevant state and federal agencies. In the event of a negative report, the five-year monitoring period begins anew and the $1,000 bond is retained until satisfactory completion. The determination of the required bond dollar amounts in each phase were based on the regulator's estimate of repair cost for the level of failure expected to occur in each phase. The regulator expected that in the initial phase any mitigation failure would be less that 100 percent and would be much lower in the second phase.

The permit for the Florida Wetlandsbank requires that the bank post performance bonds in the amount of $8,800 per acre with the city of Pembroke Pines (the landowner) prior to the commencement of mitigation work. All but $968 of the bond amounts will then be released in phases as certain milestones are reached concerning the eradication of exotic vegetation, site construction and planting, and the commencement of a five-year monitoring and maintenance period. The balance of the bond amount in each phase reflects the regulator's best estimate of the costs to repair a failure occurring in each phase. This estimate was developed, in part, from cost information provided by the banker.

Collateral Banks The performance bond approach collects funds, and only after the mitigation has not met performance standards and the banker has failed to satisfactorily correct the deficiency would the regulator move to repair the mitigation. Another option available to address controllable failure risk would be to establish a functioning wetland restoration site to serve as a "collateral bank" to secure advanced compensation. The collateral bank could be developed at public expense or might be operated under a contractual agreement between the regulator and a private party. Credit suppliers, as they sold credits from their own mitigation site, would be expected to "lease" equivalent credits from the publicly run or certified collateral bank. The amount of collateral credits that credit suppliers would be required to lease from the collateral bank would be based on the regulatory agency's estimate of the costs to create the collateral bank credits and the failure probability for the credit supplier's mitigation site. The cost to lease mitigation credits from the collateral bank should reflect credit production costs and interest charges on invested capital, including allowance for a competitive return on that capital, and would be set as follows. The cost for a credit at the collateral bank would be established and weighted by the regulator's estimate of the probability of failure at the credit supplier's mitigation site. Thus, if production costs at the collateral bank were $30,000 per credit (including interest charges) and failure probability at the private site was expected to be 50 percent (and this failure would be complete), a lease price of $15,000 would be charged.

As with the performance bond option, once the credit supplier's mitigation site was certified as successful, the lease payment would be refunded with interest. In this case, the amount refunded would be reduced by the allowance for a necessary profit if the collateral bank is privately developed. As with the performance bond, the credit supplier should be able to assess and reduce failure probability if the criteria for success are well defined by the regulatory agency.

If the credit supplier's mitigation site were judged a failure, on the other hand, then all the supplier's deposits to the collateral bank would be kept and the collateral bank would have less credits to lease. As failures occurred, the forfeited deposits would be used to create new collateral bank mitigations. In the case of failure, the required mitigation compensation would come from the collateral bank and not from the repair of the failed bank site.

Insurance The regulatory agency may alternatively choose to charge an insurance premium against controllable failure risks as a condition for selling or purchasing credits. This would be a one-time and non-refundable payment made by the credit supplier or permit applicant for each credit traded. The premiums would be collected by the regulator, placed in a fund, and used to repair or even fully replace failed mitigation sites. This option shifts failure risk costs to credit suppliers and/or permit applicants, but once the insurance payments are made, the public sector would be accepting the responsibility to ensure that wetland restorations or mitigation repairs were made to offset project failures. The premium would be based on an actuarial analysis of the probability and cost of project failure.

Such an insurance premium is required by the draft guidelines developed by Placer County in California for the establishment and use of commercial mitigation banks to provide compensation for wetland impacts that fall outside federal regulatory jurisdiction. The guidelines stipulate that credit purchasers must pay an additional 25 percent of credit costs to the county, which shall be held in a reserve account in order to provide for any remedial measures that might be necessary at commercial banks or to provide replacement wetlands at some other location. The 25 percent figure represents the county's assumptions regarding expected failure probability and repair cost, taking into consideration the other bank requirements imposed by the guidelines. However, the 25 percent figure must be considered somewhat arbitrary since it is necessarily divorced from the specific circumstances of failure probability and repair cost at particular bank parcels.

The previous discussion illustrates the potential range of mechanisms that could be included in the contracts for private commercial banks (or wetland development permits) to allocate the risk costs of mitigation failure resulting from nonperformance with contract requirements. These liability rules should be viewed as substitutes for each other, and their use could vary by situation. Moreover, the level of risk cost established for any particular bank must reflect realistic failure probability and repair cost for that bank.

The need to adjust liability rules according to the previously listed six factors, which bear on failure probability and repair cost, underscores the argument that the potential of private credit market systems requires balancing the set of trading rules imposed on any particular bank. In the extreme, the specific trading rules and bank circumstances underlying the six factors, particularly that for the timing of credit marketability, might be so stringent and favorable for mitigation success that financial assurance becomes unnecessary.

This trade-off is illustrated by the permit for the Florida Wetlandsbank as well as the draft MOA for a proposed private commercial bank in Virginia that appears to be nearing final regulatory approval. While the Florida Wetlandsbank is permitted to sell credits concurrently with the construction of mitigation parcels for those credits, the bank also intends to provide some advanced mitigation (i.e., in place and functioning). The bank's permit specifies that the performance bond requirement for concurrent mitigations is waived in the case of mitigation parcels constructed in advance of credit sales.

Credit Valuation and Trading

Central to wetlands mitigation is the need to evaluate and express the ecological worth of replacement wetlands in measures of mitigation credits and the need to determine the number of credits necessary for any bank trade to provide the required compensatory mitigation for permitted wetland impacts. The first need can be termed "credit valuation" (or crediting), while the second can be termed "credit trading" (or debiting).

Key conceptual issues relating to these needs are discussed later. While the discussion focuses on commercial mitigation banking, the issues and problems for credit valuation and trading in this context are no different than those encountered by other mitigation options (i.e., project-specific mitigation and single-user banking) which have been used extensively for over a decade. Therefore, the challenge of credit valuation and trading poses no unique problems for commercial mitigation banking nor do these needs impose a critical barrier to the widespread emergence of the private credit market alternative.

A mitigation credit is a unit measure of the increase in wetland functional value achieved at a wetland mitigation site. Mitigation credits serve as the unit of exchange for the provision of compensatory mitigation. Protocols to assess the functional value of bank replacement wetlands, as well as to measure functional losses at the permitted site, are critical for determining the acceptability of any bank trade. Without such protocols the appropriate credit requirements for a bank trade cannot be evaluated, and therefore, it is not possible to be confident that regulatory goals will be achieved through credit trading.

As the Section 404 program has grown, advances in the sophistication of methods for wetlands functional assessment have followed. However, the state-of-the-art in wetlands assessment is still experimental and somewhat controversial. Wetland functions are notoriously difficult to measure individually or cumulatively in any qualitative or quantitative way, and there is no one generalized or "correct" assessment methodology that is applicable to all wetland types and landscape settings. Nevertheless, the existing mitigation experience shows that creative ways can and have been found to directly or indirectly assess wetland functions in order to satisfactorily perform the credit valuation task.

Theory and practice suggest that the primary guiding principle for the development of credit valuation protocols for any bank relates to the needs and goals of the applicable watershed (as determined by resource managers and regulators) and the specific ways in which the bank intends to contribute to their achievement. Since watershed goals vary from area to area, and the specific ecological objectives of banks vary from bank to bank, one would expect each commercial bank to have its own unique credit valuation protocol tailored to the wetland functional values of interest in the watershed.

For example, if watershed goals focus on a suite of wetland functions, then a bank might build its credit valuation protocol around an assessment method capable of evaluating such a range of functions (e.g., the "Wetland Evaluation Technique"). If, on the other hand, watershed needs focus primarily on wildlife habitat, this might dictate the use of a narrowly defined assessment method based on that wetland function (e.g., the "Habitat Evaluation Procedures"). Both approaches are useful for evaluating credit trades involving like wetland types and might also be tailored to evaluate trades involving dissimilar wetlands when such out-of-kind trading would contribute to watershed goals. Alternatively, if watershed needs dictate in-kind trading, bank credit valuation might be based on a more simplified method for subjectively scoring acres of like wetland types.

Another consideration for the development of credit valuation protocols relates to the difficulty and expense of applying direct functional assessment methods. In general, the more technically sophisticated and comprehensive the functional assessment method used, the greater will be the cost and complexity of the credit valuation task. Since the precision of wetland functional assessments do not necessarily move in lockstep with the degree of methodological sophistication, banks often choose to focus on in-kind trading of like wetland types to facilitate the use of more simplified assessment approaches for credit valuation.

The need for banks to establish cost-effective credit valuation protocols based on watershed- and bank-specific mitigation goals means that there are as many ways in which credit valuation can proceed as there are different banks. And since credit valuation protocols will vary across banks, so will the units in which credits are defined (i.e., the credit "currency"). This is because credit currency is largely determined by the functional assessment method used for credit valuation. Depending on the assessment method used, bank credits might be defined in terms of, for example, some integrated index of wetland functioning, habitat units, or acres of like wetland types.

While credit valuation protocols and credit currency will vary from bank to bank, there must be consistency in the way credits are evaluated and defined across all uses of any particular mitigation bank. That is, the application of a credit valuation protocol to evaluate and express the ecological value of bank replacement wetlands in mitigation credits will determine the baseline methodology and currency in which all trades from that bank should be evaluated.

Once the ecological value of permitted wetland impacts and bank replacement wetlands have been assessed in the same manner and measured in the same credit currency, it then must be determined how many bank credits will be needed to provide the required compensatory mitigation for the permitted impacts. The terms by which bank credits are traded for units of permitted wetland loss can be termed the "trading ratio" (compensation or replacement ratio). Key issues relating to the determination of the trading ratio for any particular bank trade include questions relating to who should make this determination as well as how and when it should be made.

It is the responsibility of regulators to determine the trading ratio required for any particular use of bank credits as compensatory mitigation in order to ensure that regulatory goals are achieved. The presumption is that regulators will make this decision for each fill permit proposing to use a bank so as to ensure that, at a minimum, mitigation trades result in no net loss in the long-term functioning of wetlands in the applicable watershed. In other words, baseline trading ratios should be at least 1:1 for mitigation credits defined in terms of wetland functioning.

To illustrate the 1:1 trading ratio, consider the following hypothetical example. Assume that a permitted development project will result in the unavoidable loss of one wetland acre, and the permittee decides to pay a commercial mitigation bank to provide the required compensatory mitigation. Assume further that regulators, using the bank credit valuation protocol, determine that the impacted wetland has twice as much ecological value as that of the bank wetland. The 1:1 trading ratio for credits defined in terms of functional units, when translated into area requirements for compensatory mitigation, would thus require two acres of bank wetlands as compensation for the one acre of permitted wetland loss.

Regulators may, however, make the trading ratio for any bank trade higher than 1:1 for three possible reasons. For example, the trading ratio might be adjusted upward to account for the risk of mitigation failure. The use of trading ratios for this purpose was discussed earlier as one among several possible ways in which regulators might insure against the risk of mitigation failure.

A second reason why regulators may want to adjust trading ratios upward involves possible temporal losses in wetland functioning between the time at which bank wetlands are used as the basis for credit trades and the time at which these wetlands reach functional maturity. A higher trading ratio for this purpose would thus trade off less than equivalent functional value in the short term for the opportunity to obtain a net gain in wetland functioning in the long term.

Finally, regulators may want to adjust trading ratios upward to ensure that bank trades result in no net loss in wetland acreage as well as function. For example, it is possible that a bank trade based on a 1:1 trading ratio for credits (as defined in terms of units of wetland functioning) could result in a net loss in wetland acreage while at the same time ensuring functional equivalency. This could happen if the bank wetlands were judged to have greater ecological value than the impacted wetlands, so that when the 1:1 trading ratio for credits was translated into area requirements for compensatory mitigation, less than one acre of bank wetlands would be required for every one acre of permitted wetland impact. In such cases regulators may choose to adjust the trading ratio upward to ensure no net loss in both wetland acreage and function. Moreover, even in cases in which a 1:1 trading ratio would ensure a no net loss in both function and area, regulators might dictate higher than 1:1 trading ratios if they sought to achieve net gains in wetland function or acreage through bank trades.

The final consideration for the determination of trading ratios relates to when this determination should be made. As long as trading ratios are based on credits defined in terms of functional units, regulators can state up front that all credit trades involving a particular bank would be exchanged on a 1:1 basis, or some higher basis to account for risk, temporal concerns, or a net-gain objective. Then for each proposed bank trade regulators could determine the areal mitigation requirements that would achieve functional equivalency. However, in the commercial banking context it does not make any sense to define up front a set trading ratio for all bank trades if ratios are defined in terms of acres rather than credits measuring wetland functioning. This is because the particular wetland impacts to be compensated for through a commercial bank will not be known in advance of trades.

The two private commercial banks that had received federal permits as of the start of 1993 illustrate different approaches to the credit valuation and trading tasks. The permit for the Millhaven Plantation bank defines credits in terms of acres of wetland type, and gives the Corps project manager authority to make final determinations of the number of credits generated by restored bank parcels after the relevant resource agencies have had the opportunity to review and comment on the quality of the restoration work. Further, the Corps project manager has sole discretion to assess the relative functional values of replacement wetlands and impacted wetlands for the purpose of determining the trading ratio required for any particular trade. In making this determination, "the Corps may use any available technology, resource or information it determines appropriate in performing these assessments and making wetlands functions and values determinations." Trading ratios will depend on a number of factors, including the particular types of impacted and replacement wetlands (out-of-kind trades are acceptable), their relative maturity, and the nature and level of their ecological functioning.

The permit for the Florida Wetlandsbank (FWB) specifies a much different approach for credit definition and evaluation. Credits are defined in terms of "integrated functional units" based on a functional assessment methodology developed by the Corps and EPA for everglade-type wetlands. This method evaluates wetland pollution assimilation, habitat, and flood control functions and translates these assessments into a single "integrated functional index" (IFI) value. The permit specifies that the FWB mitigations will result in a specific IFI value, which "takes into consideration that the proposed bank represents and will function as a stand-alone system which will provide water quality, habitat and flood flow attenuation functions." To determine the amount of replacement wetlands required for any particular trade, an IFI value will be assessed for the impacted wetland and then translated into "FWB equivalent mitigation acreage."

Regulatory Rules to Facilitate Private Credit Markets

Both trading and regulatory rules influence a permit applicants' demand (willingness to pay) for credits and private commercial bankers' supply of credits (willingness to make investments in credit creation). Rules for the timing of credit sales, standards of performance, and liability for project failure will influence entrepreneurs' willingness to invest in supplying credits. However, for the full potential of the credit market to be realized, the demand for credits must be assured and the prices received for credits must be adequate to earn a competitive return on the investment in credit creation. Regulatory rules to promote these results should (1) facilitate market entry opportunities for private commercial banks and (2) integrate mitigation banking into watershed planning and management.

The benefits of private credit market systems would be enhanced if a sufficient number of private credit supply firms enter the market, making the supply of credits adequate for mitigation needs. Also, if there were many firms, competitive pressures would encourage those firms to continuously seek ways to lower costs. Of course, the general market conditions must be favorable for market entry to occur. For example, private banking would not be profitable in locations where there is little demand for wetlands development permits. However, even where there is a strong potential demand for credits, regulatory rules must encourage market entry by avoiding actions that inadvertently reduce the demand for credits. There are four areas for attention.

Consistency in Mitigation Requirements The demand for credits supplied by private commercial banks will be reduced if the regulatory process does not hold on-site mitigations to comparable standards as those applied to bank mitigation projects. For example, in the past some single-user banks have not been allowed to withdraw credits until the bank mitigations were in place and certified as fully successful. Only then would wetland development permits be issued in return for compensatory mitigation from the bank. This requirement discourages banking of any type and encourages permit applicants to propose on-site mitigation, which is not held to advance mitigation requirements. At the same time, the implementation and enforcement of quality standards for on-site mitigation has been lax. Indeed, it has been the failure of on-site mitigation that has promoted interest in banking.

If this inconsistency in requirements for on-site mitigation and banking continues, then some permit applicants will be encouraged to choose the apparently "cheaper" alternative of on-site mitigation (despite the likelihood of failure) and seriously dampen the demand for private bank credits. Consequently, there needs to be across-the-board regulatory reform to assure that requirements for success are the same whether the mitigation is on-site or through a bank. Private bankers are concerned about the possibility that they will be held to higher standards than those who mitigate on-site. The entrepreneurs behind the Millhaven bank suggested that this was a primary concern about their potential for financial success.

Competition from Public Banks The emergence of private credit markets may come slowly, although there is significant interest and activity among entrepreneurs. In the interim, regulators may develop a banking system that brings public commercial banks into the supply side of a mitigation credit market. There are a number of potential barriers to bringing the public sector into the mitigation supply business. However, one major problem noted earlier involves the lack of public funds for financing the construction of public commercial banks. This problem may also plague fee-based mitigation systems that collect fees in advance of the provision of mitigation, since there is no guarantee that dedicated fee revenues will actually be used for this purpose. Still, there are dozens of operating and proposed public commercial banks and fee-based mitigation systems.

Under a public credit market system, the regulatory agency is responsible for producing wetland mitigation credits and recovers production costs through the sale of credits. However, unless public banks set credit prices (or in-lieu fees) at levels that recover all mitigation costs, including interest charges on invested capital and failure risk costs, they will have a competitive price advantage over private commercial banks. If the price-setting process for public banks does not reflect all bank costs, then public banks will not only directly subsidize the mitigation of permit applicants but also will introduce "below-cost" competition for private banks. This would cause the same problem for private banks as that produced by competition from lax regulatory standards for on-site mitigation. This does not mean that public banks should set prices as high as private banks in all cases, however. Due to particular circumstances, a public bank may realize some scale economies or lower failure risk costs. If this were the case then such efficiencies would justify a lower public price than private price.

Also, many of those interested in investing in private mitigation banks question whether public entities could adequately assess the financial risks of public bank ventures. The Bracut Marsh public commercial bank developed by the California Coastal Conservancy illustrates this problem. Although operational, the bank has failed to be self-supporting, and the conservancy forecasts that when all bank credits have been sold at proscribed credit prices the bank will have recovered only 54 percent of total costs (see Environmental Law Institute, 1993).

Regulation of Private Credit Prices Compensatory mitigation requirements (and other mitigation sequencing rules) put a "mitigation price" on receiving a wetland development permit. In the same manner, private markets in mitigation credits would put prices on permits. Once the trading ratio was set for a particular trade, the permit applicant would seek credits on the open market. The price per credit in that market, times the number of credits required to satisfy mitigation requirements, would establish the price for the permit.

Consider the following hypothetical situation. A private credit supplier can produce each credit for $5,000. At the same time, a permit applicant who stands to make a profit by developing a particular wetland site is willing and able to pay as much as $50,000 to obtain the permit. During the regulatory review process the regulator considers failure risk and determines that the permit will be granted if the applicant provides three units of mitigation (i.e., credits) for the one unit of wetland function lost due to the development project (or 3:1 trading ratio). Knowing this ratio the permit applicant begins a negotiation with the credit supplier.

One possible outcome is that the permit applicant will only pay the credit supplier a competitive return price of $5,000 per credit, incurring a total cost of $15,000 for the permit. A $35,000 development surplus would then remain with the permit applicant. Another possibility is that the supplier is the only one in the area certified by the regulator (the seller has been sanctioned as a monopoly) and is able to extract the full $50,000 of the permit applicant's willingness to pay. In this case the $35,000 development surplus has been transferred from the permit applicant to the credit supplier. In either case the secured replacement in wetland function is unaffected—the ratio is 3:1.

There is a third possibility. Suppose that before setting the trading ratio the regulator knew the permit applicant's willingness to pay ($50,000) and the credit supplier's minimum price for selling each credit ($5,000). In this case the trading ratio could be set at 10:1 and a deal between the applicant and credit supplier might still be made. In this case, the $35,000 of development surplus would be transferred to the wetland resource or, more generally, to the public.

One perspective on these different distributional outcomes might be that the permit applicant has a property right to the site and its value. If the public is satisfied with the 3:1 compensation level, and if the credit supplier earns a return sufficient to keep resources in the mitigation supply business, then the $35,000 should stay with the applicant. Such a view might call for price controls of some sort on the market if there is little price competition among suppliers. In fact, some regulators at the field level expressed the concern that private entrepreneurs might make "too large" a profit from selling wetlands credits; that is, prices would be "too high." While they did not advocate price controls, they instead saw this as a reason to discourage private markets in mitigation credits. These people seemed to favor public banks in part for this reason. However, this viewpoint was not held uniformly by all regulators.

Another perspective is that the only reason for the 3:1 trading ratio is that the public did not realize how much the permit applicant was willing to pay for the permit. If this willingness to pay were known by the regulator, then the net gain goal could be advanced by insisting on as much as a 10:1 trading ratio. Interestingly, some regulators described how the determination of "acceptable" compensation for a permit often was partly established by the regulator's assessment of the applicant's willingness and ability to pay for compensation. However, offices of federal and state agencies indicated that the regulator's job was only to secure acceptable mitigation compensation for granting the permit and that the financial capability of the applicant should not be a consideration.

It may appear that one way to stimulate market entry would be for the regulator to seek a very high (e.g., 10:1) trading ratio, presumably to stimulate credit demand. However, the nature of the feedback links between the markets for permits and credits complicates reaching such a straightforward conclusion. The trading ratio and the trading rules that affect credit price together determine the price of permits. Thus, higher trading ratios would increase permit prices, blunting permit demand and then credit demand. The net effect of these countervailing forces on private banks' credit demand as trading ratios are increased would depend on general market conditions that influence the demand for wetland development.

The distribution of returns that best serves the interests of advancing the private credit market is to avoid any interference in the establishment of the price of credits and to set trading rules according to environmental criteria. If there were excess profits in private banking, that would act as a short-term and powerful incentive for others to enter the credit supply business. Expanded competition in that business might be necessary if an adequate number of credits are to be supplied through private banks in the long term. To stimulate competition the regulator should simply set trading rules and trading ratios, which satisfy environmental concerns for project failure, and then let the applicant and supplier bargain over credit prices. The regulators should also allow permit applicants to choose the suppliers they wish to deal with. In the example just cited, some return above the credit supplier's $5,000 competitive return might be extracted from permit applicants.

Market Area Definition Using ecological arguments, regulators feel that mitigation bank sites should be as close as possible to the permitted wetland. As a result, for the few private banks currently allowed to sell credits the regulators expressed the need to closely define the geographic area within which credits could be sold. In fact, an ecological basis for determining the trading area need not be set in advance of the establishment of the bank. Instead the trading area might be determined when evaluating each permit application. While in some cases there may be an ecological basis for limiting the geographic area for credit sales, generally narrowing the market area will shift (lower) the demand for credits for any single bank and restrict the possibility that numerous banks will be able to compete to serve any one area.

Another geographic factor that can shift credit demand are the criteria for wetland delineation and for program jurisdiction. Guidelines on these matters define the size of the areas subject to regulation and can affect the demand for permits and then credits. The greater the geographical extent of areas falling within the wetlands regulatory net, the greater the extent of wetland development subject to mitigation requirements. Then, as the scope of mitigation needs expands, the demand for credits at any given price would be expected to increase. While the policy decisions that could expand or contract the geographical area subject to regulation should not be based on creating market opportunities for private commercial banking, nonetheless it should be recognized that such decisions would affect credit demand.

Conclusion

Private mitigation credit markets could help the federal wetland regulatory program achieve no-net-loss of wetlands by increasing the opportunity to obtain successful compensatory mitigation for permitted wetland losses. These markets, which rely on the emergence of private commercial mitigation banks, could promote this result in two ways. First, credit markets would provide the means to secure mitigation for the many small wetland impacts that would otherwise go unmitigated. Second, the use of private credit market systems as a substitute for on-site mitigation in certain cases could enable regulators to avoid the several sources of failure associated with the on-site mitigation option.

The private market alternative is the next step beyond the recent intense interest in traditional, "single-user" mitigation banking arrangements. Private credit markets, if carefully structured, can offer a competitive return on investment for credit suppliers and an expedited permit review process for qualifying wetland developers. Most importantly, credit trading systems that insure against the risk of mitigation failure would benefit the public by advancing achievement of the no-net-loss and net gain regulatory goals.

All the various stakeholders in wetland regulation seem to agree that compensatory mitigation is not working well in practice and that the time is ripe for reform. Practical evidence of the desire for change is provided by the two newly permitted and the dozen or so emerging private commercial banks across the country. At the policy level, some states and localities have already passed legislation authorizing private credit markets and are currently struggling with developing regulations for their establishment and use. While federal government policy has not motivated these developments, recent proposals for policy reform in both the executive and legislative branches support the general concept of mitigation banking.

At this point, the widespread emergence of private credit markets depends to a large extent on policy guidance, which clarifies what is expected of entrepreneurs regarding the establishment and use of private commercial mitigation banks. Clarification of the types of bank conditions and trading rules expected under the regulatory program would eliminate much of the uncertainty currently facing prospective credit suppliers. Meanwhile, changes to the regulatory rules might also be part of an effort to facilitate private banking.

The Corps and EPA jointly issued interim guidance to help guide field staff in the negotiation of mitigation banks in the summer of 1993. This was followed by final guidance, which was proposed in March 1995. The agencies need to focus attention on the specific needs of commercial mitigation banks, which pose somewhat different challenges than single-user banking arrangements.

National guidance should give individual regions the flexibility to produce regional guidelines specific to their own watershed needs, which in turn would increase the certainty for prospective credit suppliers in those areas. Of course, regulators in the field must provide information to prospective credit suppliers on the general process and steps required for the regulatory approval of private commercial mitigation banks and credit sales.

Note

1. "Federal Guidance for the Establishment, Use and Operation of Mitigation Banks," U.S. Department of Defense, Environmental Protection Agency, Department of Interior, Department of Agriculture, Department of Commerce, Federal Register, p. 12291, March 6, 1995.

7

Legal Considerations

Lindell L. Marsh, Robert D. Sokolove,
and Robert M. Rhodes

The primary objective of mitigation banks is to address the many deficiencies of the traditional project-by-project, fragmented approach to mitigation. This approach has resulted in costly and largely unsuccessful efforts to offset impacts to wetlands. More specifically, the objectives of mitigation banks include

- provision for mitigation in advance of impacts and assurances that the mitigation will occur;
- provision for an efficient and convenient way to satisfy regulation-related mitigation requirements regarding individual development projects; and
- coordination and aggregation of the mitigation mandated for various projects into a more effective form.

The general idea is that a "banker" acquires land, creates or enhances a wetland, and establishes an arrangement for the perpetual conservation of the resulting bank. In so doing, the banker earns credits from the regulatory agencies charged with protecting the particular type of habitat or wetlands in question. As developers are required to mitigate pursuant to public sector regulations, rather than undertake their own mitigation projects, they purchase mitigation credits from the banker. Unfortunately, however, mitigation banking has met with limited success. Yet, it continues to make sense, provided that certain assurances can be provided to the various interests involved. These include the following:

1. Regulatory agencies want assurances that the wetlands or wildlife habitat proposed will be created, enhanced, and maintained in perpetuity and will fully mitigate the development-related impacts as required by applicable regulations.
2. The banker wants assurances that the requirements of the agencies are clear and attainable and that a market is available for the resulting credits.
3. Developers want predictability—certainty, regarding the cost and effectiveness of the credits in satisfying the particular regulatory mitigation requirements. Ideally, a developer would like to simply pay a fee and have no further responsibility.

The model response that has been suggested to satisfy these desires is for the regulatory agency or agencies to enter into an agreement with a mitigation banker, whereby the banker undertakes to create the bank and improve and/or conserve the habitat within the bank site in perpetuity. In exchange, the agencies commit to honor credits for the mitigation provided by the bank, in an agreed upon amount, as mitigation for impacts from other projects on the type of wetlands or habitat of concern. This simplified model, however, leaves a number of questions unanswered. For example:

- How is a banker assured that agreements will be honored by the agencies, in light of the complexity of the science involved and changes in agency staffs over time?
- How is a bank sponsor to be assured that once a bank is created, the regulatory agency will not encourage prospective purchasers of credits to look elsewhere for mitigation?
- Is there a way for the credits to be available before the bank is fully completed?
- What are the alternative forms for the long-term ownership, funding, and operation of the bank? Must the bank site be held in fee title or can a conservation easement be used? Should the bank be owned long term by the banker, a nonprofit organization, or a governmental agency? Should funding be by up-front endowment, assessments on adjacent development, or tax revenues?

To some extent, as with most transactions and relationships, the answers to these questions cannot fully be provided by legal provisions or assurances. For example, while a banking agreement can provide for fairness in making a determination, arbitration if there is any disagreement, and covenants on the part of the agencies that prospective purchasers will not be encouraged to go elsewhere, it may be difficult to invoke and enforce these provisions. Establishing a history and embedding the principles of this approach into the culture of the organizations involved is critical to the effectiveness of the concept.

Permits, Agreements, and Plans

The regulatory permit issued to a developer or mitigation credit purchaser, establishes the mitigation requirement that must be met for a specific project.[1] The mitigation bank simply provides a convenient source for that mitigation through the purchase of credits, as an alternative to the developer performing the mitigation itself. The use of a mitigation bank does not modify the mitigation requirements that the party must satisfy. The use of a bank, however, should result not only in a more convenient source of mitigation but also in less costly mitigation. For example, where the bank has created new wetlands that are fully mature at the time mitigation is required, a lower mitigation ratio should be applied by the regulatory agency to reflect the removal of uncertainty concerning the viability of the mitigation effort.

Further, in addition to the permitting requirements in connection with the development impact, the creation and operation of the bank itself may also require permits. The institutional arrangements and agreements establishing the bank are created in connection with these permits. For example, for a bank in California that will affect wetlands within a nonnavigable streambed, to create a bank the bank sponsor will need, at a minimum, local approvals (general plan, zoning, and grading and building permits), a state streambed alteration agreement with the Department of Fish and Game, and a federal Clean Water Act Section 404 permit. If the bank involves coastal wetlands, permits or other approvals are required to be obtained as well from the California Coastal Commission under the California Coastal Act.

Also, the bank sponsor will want to be assured, before it undertakes development of the bank, that the agencies will accept the credits created by the banking scheme as mitigation for impacts (debits) by other developers to whom the bank sponsor hopes to sell such credits. If the bank sponsor does not have these assurances before it creates, restores, or enhances the bank's wetlands, it will be at significant risk that these agencies may not thereafter honor the credits created or grant permits to allow it to undo what it has done.

Historically, agencies have undertaken to provide these assurances in a variety of ways. In some cases, "handshake" agreements have been used, albeit with extreme risk. An example is the Louisiana Department of Transportation and Development Mitigation Bank where the lack of a written agreement gave rise to disputes and significant delays.[2] At the other extreme are agreements accompanying highly detailed plans, such as the West Eugene Wetlands Plan and Mitigation Bank, discussed in the Case Studies section following the chapters, which not only provide the specifics regarding the bank but also fits the bank into the larger context of regional wetlands management.

Currently, virtually every existing and proposed bank employs some type of formal instrument that memorializes the terms under which the bank will operate. These instruments may take a variety of forms, including, for example: (1) a memorandum of agreement or understanding, (2) an individual Section 404 "fill" permit, (3) an individual Section 404 "banking" permit, (4) a Section 404 general permit, (5) a corporate charter, and (6) legislation or regulation. The most common instruments are the memorandum of understanding (MOU) or memorandum of agreement (MOA) between the permitting agencies and the credit producer. More than 50 existing and proposed banks have final or draft MOAs or MOUs.[3] In addition, other proposed banks have indicated that one would be drafted later in the planning process. Alternatively, the permitting agencies may simply ratify the credit producer's proposed site plan through formal letters of assent in lieu of a separate MOA, as was done with the Patrick Lake bank in Wisconsin.

There may be little distinction between an MOU and MOA. The term MOU, however, tends to raise questions regarding the enforceability of the instrument, as discussed in greater detail later. Accordingly, the term MOA is used to indicate that the instrument should be enforceable.

MOAs make sense for a number of reasons:

- A permit evidences the granting of permission by the regulatory agency to the permittee to undertake certain otherwise prohibited actions subject to compliance with specified conditions. Generally, permits tend to be issued by each agency separately. In some cases they do not even require the signature of the permittee. Their primary enforcement device is revocation. In contrast, an MOA contemplates the evidencing of a broader set of mutual covenants, as well as conditions. Further, MOAs may coordinate and reconcile compliance with the regulatory requirements of several agencies, as well as private sector interests. In some cases, it may provide for certain commitments by one agency to another in the implementation of the bank. It is a very flexible and inclusive instrument.
- Often a permit is for a limited term. For example, a federal Section 404 permit lasts for no longer than five years.[4] Accordingly, the undertakings and

assurances set forth in the MOA should extend for a longer period of time and more rigorously address the circumstances under which they can be terminated. Implementation agreements developed under Section 10(a) of the federal Endangered Species Act (ESA) regarding habitat conservation planning, provide a good model. These often provide for a term of 30 or more years and specify the conditions under which they may be terminated.

- It is preferable for the agreed undertakings of the regulatory agency, for example, with respect to the provision of credit in the future, to be articulated as covenants in an agreement rather than in a permit that provides for the granting of permission with various attendant conditions. Often the result of an attempt to provide for such a covenant in a permit alone is of questionable enforceability unless carefully and properly drafted.
- From the perspective of the regulatory agency, once the credits are sold and the initial work is done to establish the bank, revocation of the bank permit is of little value to assure the acquisition of the agency's objectives. Further, where credits have been conveyed to a number of developers, it would be unfair to terminate the permits of all such developers where the default perhaps only relates to one of them or to the bank operator. The MOA format can provide more effectively for a flexible range of remedies to be available to the agency.
- An MOA can provide more effectively for various circumstances that must be collaboratively addressed concerning the long-term operation of the bank. For example, it may be desirable to provide for more adaptive management of the wetland to be conserved. This can be addressed better by procedures set forth in an MOA than in permit documents.

In summary, as suggested by the Federal Guidance, probably the best approach is to use a bank agreement or MOA, evidencing the agreement of the involved local, state, and federal agencies with the bank sponsor.[5] The process required to obtain the approval of such an MOA, however, may be complex, difficult, and time-consuming.

The agreement document, whether it is an individual permit or an MOA, is often accompanied by a plan. The plan may apply to a specific bank or it may cover a watershed, habitat, range of habitats, or a geographic area defined as such for other purposes (e.g., the Lake Tahoe region). It may cover a few acres or thousands of acres. It may be named a "banking," "conservation," "resource management," "special area," or "habitat conservation" plan depending upon the regulatory context and other circumstances. It is usually prepared by a biologist or planner and is normally reviewed by an attorney. From a legal perspective, a plan should clearly, unambiguously, and accurately reflect the agreement of the parties. Indeed, the regulatory agencies and conservation interests often insist that compliance with the plan be made an express condition of the respective permits. One measure that may be taken to lessen the possibility of conflict is to stipulate in the MOA that the plan was not prepared by attorneys nor intended to be a legal document. Further, the MOA may indicate that any conflicts between the plan and the MOA shall be resolved in favor of the MOA. Obviously this is not as satisfactory as a plan that is crafted with the same care as a legal document. However, given the number of different professionals and interests involved, it is very difficult to achieve such a high level of legal clarity in an expeditious and cost-efficient manner.

While the permits, banking agreement, and plan evidence the central assurances required by the various interests as outlined earlier, they are normally accompanied by a broader suite of legal documents. These documents typically include a bank ledger maintained by the banker and the lead agency, endorsements of credit availability, a conveyance of the bank lands, agreements as to the operation and maintenance of the bank, and a conveyance of a conservation easement restricting the future uses that can be made of the bank lands. Interestingly, the Homebuilders Association of Greater Chicago has proposed that further assurances regarding the limits on use of the bank be provided by a special state corporate charter to be required of any bank operator.[6] This charter appears similar to that required by some states of cemetery operators. It would limit the resultant corporation's powers to sell, convey, or encumber the bank lands. Another approach for completing the suite of assurances is reflected in the Oregon Mitigation Bank Act.[7] This act authorizes the Director of State Lands to create up to four pilot mitigation banks. In so doing the director is to set forth in detail an administrative framework governing their operation. Another example is the ordinance adopted by the city and borough of Juneau, Alaska. It is patterned on the Oregon statute and governs the public mitigation banking called for by its Wetlands Mitigation Plan. Along with establishing substantive policies, the Juneau ordinance created a Mitigation Banking Board. This board is authorized to "adopt . . . standards and criteria for the site selection process, operation and evaluation of mitigation banks."[8] When utilizing this type of legislation-based approach, any mitigation bank would be required to comply precisely with the provisions laid out in the enabling statute. This is not dissimilar to the approach used in civil law countries whereby many of the provisions determined by agreement in this country are mandated in those countries by statute. The advantage of such a statutory scheme is that it simplifies the MOA. The disadvantage is that it significantly restricts the flexibility otherwise provided by an MOA.

The specific elements of the framework of assurances created by this suite of legal instruments and statutes are discussed in greater detail herein.

Terms of the Banking Agreement

Under usual circumstances a banking agreement should address the following topics, which are subsequently discussed in greater detail:

- Parties
- Title of the Bank Lands
- Goals and Objectives of Bank
- Service Area: Potential Credit Purchasers
- Determination of Credits
- Record Keeping and Certification of Credits
- Availability of Credits
- Implementation Program
- Dedication of Use
- Management and Operation
- Long-Term Budget and Funding
- Monitoring and Reporting
- Assignment and Delegation
- Remedial Action: Extraordinary and Unforeseen Events
- Remedies
- Term
- Miscellaneous Provisions

Parties

The banking agreement should be viewed as a free-standing document, the performance of which can be made a condition of the various required permits and approvals of the involved local, state, and federal agencies. Ideally, the bank sponsor and each of these agencies should be a party to the agreement. However, the historic practice has been for each agency to require a separate permit and, often, agreement. The agreements entered into under Section 10(a) of the ESA provide a model for a multiagency agreement that can act as a common component of the various agency permits and approvals.[9] The San Bruno

Mountain, California Habitat Conservation Plan (HCP) was signed by three cities, a county, two state agencies, and the United States Fish and Wildlife Service (USFWS). (No wetlands were involved in the plan.)[10]

While it is desirable to have a single agreement, it may be that this is so foreign to the customs or processes of one or more of the agencies that the bank sponsor may choose, or be forced to choose, to utilize separate agreements. The problem, of course, is that the terms of the various agreements may not be consistent: an early commitment to one agency may not be compatible with the later demands of the other agencies.

Title of Banked Lands

The regulatory agencies must be assured at the time the banking credits are available that: (1) the lands will be used for the agreed-upon conservation purposes, normally in perpetuity, and (2) they have the right to take remedial action if necessary. These objectives can be accomplished in a variety of ways. For example, it may be sufficient at the outset of the project for the bank sponsor to have a mere option to acquire fee title or an easement with respect to the banked lands. In fact, if the underlying owner is willing, the bank sponsor could undertake work on the lands prior to obtaining title. However, prior to the availability of the banking credits, fee title or an adequate easement must be obtained by the bank sponsor. Further, the fee or easement held by the bank sponsor must be free and clear of liens, easements, encumbrances, and other interests that would impair the use of the lands for bank purposes. Assurances of title can be provided by warranties and representations in the banking agreement, title insurance, or a title opinion.

There are several ways in which the bank sponsor or bank operator can provide assurances to the regulatory agencies that the bank lands will be operated only for bank purposes. The bank sponsor or bank operator can convey fee title or a conservation easement to one or more of the regulatory agencies or to an agreed upon nonprofit conservation organization. Alternatively, covenants running with and encumbering the bank lands can be used. However, their enforcement is subject to certain equitable doctrines (e.g., change of circumstances). Thus, they do not provide the same level of protection as the conveyance of a conservation easement. In connection with the conveyance of a conservation easement or a real covenant, assurances regarding title should also be provided to the receiving agency.

In the event that a conservation easement is used to protect the bank lands, it should include the right to enter them and remedy any violation of the banking agreement. If fee title is conveyed to an agency or conservation organization, provision must be made for funding the management and operation of the banked lands. In some cases, the agency or organization may undertake the operation of the banked lands, utilizing general tax revenues or assessments on development within the area. In the San Bruno Mountain HCP, a charge was to be levied on all of the lands developed by the developer.[11] For the large-scale Natural Community Conservation Program (NCCP) plans in southern California, an assessment on development is being contemplated. However, on a banking site–by–banking site basis, this is often accomplished by use of an endowment fund. The earnings of such a fund, when invested, provides sufficient income for the management and operation of the banked lands.

Goals and Objectives of Bank

It is important that the banking agreement set forth the goals and objectives of the bank in detail. The banking agreement normally will reflect the agreement of a number of agencies and interests and may address a very complex ecosystem or group of resources. In such a context it may be very difficult to establish unambiguous benchmarks, standards of performance, and conditions. The more detailed the description of existing conditions, conditions of performance, and alternatives, the better.

Service Area: Potential Credit Purchasers

The agreement should define the class of impacts, or debits, to which the credits can be applied. Often, the class is defined in geographic terms. The Federal Guidance suggests the designation of a "service area."[12] The California Mitigation Banking legislation for the San Joaquin Valley establishes a 40-mile radius.[13] In other cases, the debits and credits must occur in the same watershed or within the range of a species, comprised of a particular habitat or habitats or within the boundaries of a political jurisdiction. One of the values of embedding a bank in a broader plan is that the underpinnings of the geographic parameters of the bank can be better understood. Thus, the resulting mitigation standards may be articulated in a less arbitrary manner. This should provide increased flexibility and effectiveness.

In determining the service area, a distinction may be made among different classes of developers. In southern California, utilities often are allowed to mitigate for endangered species habitat anywhere within the range of the species. For example, for the California gnatcatcher, this range extends from the Mexican border northward approximately 100 miles. It can be argued that resource agencies should be allowed greater flexibility in developing a conservation strategy in connection with regional multiple-species plans. This broadens the potential class of developers and credit purchasers that can utilize a bank. The Bank of America's Carlsbad Highlands Conservation Bank, described in the Case Studies section following the chapters, allows credits to be used to mitigate impacts anywhere within the NCCP planning area (the three counties comprising the area of southern California south of Los Angeles and north of the border with Mexico).

Determination of Credits

One of the most difficult issues to address in drafting the banking agreement is the valuation of credits in relationship to debits (impacts). From a legal perspective there are several considerations. Is a credit based on acreage or function? Should credits be available for preservation only? Should credits be available for adjacent lands or habitat of a different type? And, at what time in the habitat creation/restoration/enhancement process should credits be made available?

One approach is for the bank sponsor to provide credits to developers for specific impacts based on an ad hoc determination of the resource agencies as to the fit between the mitigation and the impacts and the impact/mitigation ratio that should apply in that case. This is a very risky alternative in that the banker undertakes to create, restore, or enhance the bank lands before the credits are determined. It allocates a great deal of discretion to the agencies, with very little bargaining power left to the bank sponsor. The circumstance is made worse by the fact that the regulators may have a difficult time reaching agreement on the credit valuation or they may discourage the use of the bank (e.g., by requiring a greater mitigation ratio, thereby lowering the value of the credits).

An important objective, therefore, is for the bank sponsor to reduce the amount of discretion given the agencies and increase the breadth of impacts that may be offset by bank credits. Several specific considerations in this regard are discussed here in greater detail.

Increased Function. In general, wetlands credits are normally based on the increase in wetlands function as a result of the creation, enhancement, or restoration of a wetland. Biological benchmarks or performance standards are established. As these benchmarks or performance standards are reached, the regulatory agencies confirm the increased credits earned and available. The various approaches for evaluating this increase in function are discussed in chapter 9.

Based on Preservation. Historically, there has been a great reluctance to allow mitigation credit simply for preserving (as opposed to creating, enhancing, or restoring) wetlands based on the notion that wetlands are already protected by regulation and that any loss of wetlands acreage should be offset by gains (i.e., no net loss of wetlands). The Federal Guidance generally would allow credit for preserving wetlands "only in exceptional circumstances,"[14] such as where the preservation is in conjunction with "restoration, creation or enhancement activities, and when it is demonstrated that the preservation will augment the functions of the restored, created or enhanced aquatic resource."[15] The guidance goes on to suggest, however, that credits "based solely on preservation should be based on the functions that would otherwise be lost or degraded if the aquatic resources were not preserved, and the timing of such loss or degradation."[16] Upland habitat banks tend to be viewed differently, perhaps in part because of the different history of resource conservation under the Clean Water Act (CWA) and the Endangered Species Act (ESA). Under the CWA, the standard of "no net loss" was promulgated. Under the ESA, however, the focus has been on the conservation of a sufficient number of individuals of a species and its habitat to preclude jeopardy to the species' continued existence. Thus the regulatory objective tends toward the conservation of a portion of the existing population of a species. It follows that under such a system, the earlier that conservation can be addressed, the lower the ratio of habitat that must be conserved versus the habitat that can be disturbed. The result of these varying regulatory goals and methods is that there has been a greater reluctance under the CWA than under the ESA to provide credit for preservation of the resource.

Inclusion of Uplands in Wetlands Banks. The Federal Guidance, focusing on wetlands banks, would not allow upland areas to be "directly counted as mitigation credits."[17] However, they would allow a functional assessment to determine the manner and extent to which such plan features "augment" the functions of "restored, created or enhanced wetlands and/or other aquatic resources," increasing the "per-unit value of the aquatic habitat in the bank."[18]

Exclusion of Credits Based on Existing Public Projects or Federal Funding. The Federal Guidance indicates that where the bank uses public lands, the credits "must be based solely on those values in the bank that are supplemental to the public program(s) already planned or in place, that is, baseline values represented by existing or already planned public programs, including preservation values, may not be counted toward bank credits." Further, federally funded wetland conservation projects cannot be used for the purpose of generating credits.

Ratios for Specific Species, Wetlands, and Conservation Plans. As discussed earlier, under HCPs and other conservation plans addressing species that are listed or may be listed in the future under the ESA, there has been much more flexibility in determining mitigation ratios. The debit/credit ratio for wetlands mitigation has tended to be a minimum of 1:1. It is likely, though, that, as the conservation of wetlands is addressed under conservation plans like Special Area Mitigation Plans (SAMPs)[19] or is analyzed, distinguished, and graded based on its wildlife value, the same flexibility experienced under the ESA will be provided. For example, in and around Houston, Texas, there are broad areas that have been characterized by some as wetlands. If the wildlife value of these lands is the criterion for evaluation, it may be determined that only a fraction of these lands need to be preserved in order to adequately protect the biodiversity of the region. If this is so, the resulting required debit/credit conservation ratio may be less than 1:1.

Resolution of Conflicts. The banking agreement should set forth a procedure for resolving disagreements and conflicts regarding the amount of credit available or required to compensate for specific impacts based on standards included within it. The alternatives for dispute resolution may range from elevation (to higher levels within the agencies) to arbitration.

Recordkeeping and Certification of Credits

As exists with real property, a system must be provided to make sure that banking credits are available for sale and that they are not sold more than once. A simplified system involves the maintenance of a ledger that indicates the credits that are available and those that have been sold. Individual certificates or other instruments signed by the agencies assure the developer/credit purchaser that the credits being purchased are indeed available and will be honored. This can also be accomplished by including in the banking permit issued to the developer the indication that the banking credits from the bank are available and have been applied. Of course, this is not effective where the banking credits are acquired, or contracted for, prior to the issuance of the permit in favor of the developer/credit purchaser. The banking agreement could include a delegation of the certification function to a single representative agency, such as the Army Corps of Engineers.

Availability of Credits

One of the most critical issues regarding the sale of wetland credits is the question of whether they can be sold before the wetlands in the bank have been established successfully. Generally, a banking agreement should provide that credits shall be available when the creation, restoration, enhancement, or preservation is certified as complete by the regulatory agencies involved. Mitigation bankers generally agree that a portion of the proposed wetland credits should be available for advanced sales in increments. A reasonable scheme may be to allow a mitigation banker to sell and convey a portion of its wetland credits upon the signing of a bank MOA, another portion when the site hydrology is established, another portion when construction is commenced, and a final portion when planting is completed and/or the wetlands are shown to be viable. In order to take advantage of these advanced sales, though, the mitigation banker should be required to post a performance bond or other form of surety equal to the cost of construction before the banker is allowed to make such sales. This will ensure construction of the bank if the banker defaults. Such a scheme can balance the financing needs of the mitigation banker with the desire of the environmental and regulatory communities to reduce the temporal loss of wetlands and alleviate their concerns regarding the likelihood of failure of the wetland restoration and creation efforts. Of course, the mitigation banker may enter in agreements with prospective credit purchasers, whereby the banker commits to convey the credits when they become available.

Implementation Program

In addition to the permits and banking agreements, it is often helpful to prepare a conservation plan, implementation program, or banking plan that separately describes the work to be done and the long-term conservation program for the banked lands from a biological perspective. The banking agreement can then implement the provisions of this plan, as appropriate. This procedure provides a useable manual for those actually engaged in performing the work that is separate from the complex legal provisions contained in the MOA. Such banking plans are functionally similar to HCPs, conservation plans, and SAMPs under the ESA and the CWA.

Dedication of Use

Assurances that the lands will be used in accordance with the banking agreement can be provided by: (1) the conveyance of a conservation easement to one of the regulatory agencies, (2) covenants that run with the land, or (3) conveyance of fee title to a public agency or nonprofit conservation organization, such as a land trust. The use of real covenants is the least effective form of assurance due to the fact that they are subject to equitable doctrines that allow a court discretion in certain instances (e.g., where there has been a change in circumstances) in choosing whether to enforce the covenant. However, they may be combined with conservation easements to a regulatory agency or conservation organization.

Management and Operation

In addition to requirements and standards regarding the creation, enhancement, and restoration of the resource, the agreement should provide for long-term management and operation. Detailed descriptions of the tasks required and the standards of performance should be specified. Provisions can be made for an annual program, budget, and reports to the agencies for their comment and approval. In some cases, a technical or management committee may be specified. However, in developing these provisions, the agencies should be realistic about their availability to participate on such committees and review and approve such programs and reports.

In some cases, the operation of the banked lands is not performed by the bank sponsor but rather by a bank operator. Accordingly, the banking agreement may provide that (1) the operator will assume the operating obligations for the benefit of the agencies and (2) the operator must be approved by the agencies (or one of the agencies on behalf of all of them). The approval of the agencies could be a prerequisite for the release of the bank sponsor from the operating obligations under the agreement. In connection with such approvals, the agreement may provide for consideration of the qualifications and financial capacity of the proposed bank operator.

Long-Term Budget and Funding

A long-term budget and the sources of funding should be provided for in the banking agreement. The budget will assist in characterizing the level of conservation anticipated. In addition, the agreement may provide for the approval of annual budgets and operating programs. However, a distinction should be made between use of a budget for establishing the funding obligation of the bank sponsor and for other purposes. For example, it may be that increases in the budget over an established minimum can be specified and approved such that they are not the obligation of the bank sponsor.

Funding can be provided in a variety of ways, including: (1) by an endowment provided by the bank sponsor; (2) by payments from credit purchasers; (3) from public sector taxes, assessments, and contributions; and (4) from homeowners' fees and charges. While endowments are most commonly thought of, other sources of funding can be specified. For example, in connection with the San Bruno Mountain HCP, new homeowners in the plan area were required to pay a $20 per year fee, adjusted for inflation, to go toward operation of the adjacent conserved habitat. In addition, under California law, a Habitat Maintenance Assessment District can be established to provide for assessments for the improvement or maintenance of natural habitat.[20]

Monitoring, Reporting, and Remedial Action

While monitoring and reporting is considered to be good practice, in some states it is also required by law.[21] The surveys and measurements required should be specified in reasonable detail and be provided for in the budgets required by the agreement. Reports should be made at appropriate times and intervals, including in connection with any annual report.

The occurrence of extraordinary and unforeseen events should be carefully addressed in the agreement. Who will bear the risk? Will the assignment of risk and level of risk change at certain points? For example, what if a flood or infestation destroys the creation of a critical vegetative element of the bank before the confirmation of the credits? What if the necessary topographic contours of the site are destroyed by flooding ten years after the confirmation of the credits? The specific provision may vary depending upon the circumstances. For example, by analogy to the payment of an insurance premium, it may be that in return for increased up-front conservation, the regulatory agencies agree that the bank sponsor will not be required to assume particular risks. The recent "no surprises" policy promulgated in August 1994 by Secretary of the Interior Bruce Babbitt in connection with Section 10(a) permits under the ESA is based on this approach.[22] That is, once an agreement is reached on the level of conservation to be provided by the applicant, there will be no further land or monetary mitigation required, even in the face of extraordinary or unforeseen circumstances. This policy reflects the need for certainty by the private sector development community, as well as the more specific requirement under Section 10(a) that the applicant avoid and mitigate the "take" of the species in question to the maximum extent practicable. Accordingly, once the applicant has met this standard, it is not required to mitigate further.

Remedies

One of the advantages of providing for a mitigation banking arrangement in an MOA form rather than in the usual permit form is that the remedies available to each party may be set forth with care and in detail. For example, it may be desirable for the regulatory agencies to expressly provide for equitable remedies (e.g., specific performance, injunctive relief, temporary restraining orders) as a consequence of the unique nature of the resource involved. Liquidated damages and damages following the completion of the work and the certification and sale of the credits may be stipulated. On the other hand, the bank sponsor will want to have equitable remedies available with respect to the obligation of the agencies to honor the credits provided and to continue issuing permits pursuant to the agreement (which may be longer than the standard five-year term of a Section 404 permit).

As discussed later, there are questions as to the extent of the enforceability of such agreements against regulatory agencies. There is no express authorization in the CWA or the ESA for the agencies to enter into agreements promising continued permit validity or the granting of new permits. However, entering into such contracts and making such commitments may be a reasonable exercise of the permitting authority of the agencies under both acts. The agreement should provide that if the bank sponsor has relied to its detriment on the representations, undertakings, and reasonable assurances of the regulatory agencies that (1) it will be allowed to continue the work, operate the bank, and provide credits to developers as compensation for their impacts; and (2) the agencies will cooperate in good faith in allowing it to market the credits to eligible developers.

Term

A term of sufficient duration should be specified in the agreement to assure the bank sponsor that it will be able to complete the bank and market the credits. Under the ESA, it is common for implementation agreements that accompany HCPs and Section 10(a) permits to have a term of 30 or more years.

MISCELLANEOUS PROVISIONS

The normal miscellaneous contractual terms should be included, such as notices, the extension of the burdens and benefits to successors and assigns, costs of enforcement, third-party beneficiaries, and so on. Consideration should be given as to whether the agreement should be recorded or whether the necessary assurances can be adequately provided by ancillary deeds, real covenants, and memoranda that are themselves recorded.

Processing Considerations

The processing of documents relating to a mitigation bank is somewhat similar to the processing of any of the permits involved, except that the institutional arrangements often contemplate agreement among several agencies and interests. Accordingly, it can be a far more complex process. From the standpoint of the bank sponsor, it is critical that the legal arrangements with any one agency are compatible with those concluded with the others. Thus, as discussed earlier, a single multiple-agency mitigation plan and agreement is preferable, with the permits and approvals of each agency being conditioned thereupon.

With regard to the requirements of the National Environmental Policy Act (NEPA) and some state analogues, there is provision for the preparation of joint state/federal documents. This can significantly simplify the process if the staffs of the various agencies are willing to work cooperatively. On the other hand, the benefit of such authorizations may be offset by the lack of coordination among agencies. In this case, it may be more expedient to prepare separate but parallel environmental reports or statements. Because of the environmentally beneficial effects of the actions contemplated, it is often possible to mitigate any anticipated environmental impacts to a level of insignificance and thereby avoid the preparation of a full environmental impact statement on the federal level. In such cases, the federal agency would prepare an environmental assessment and a finding of no significant impact.

Enforceability of the Banking Agreements

It is unclear that a regulatory agency can agree in advance to recognize the use of particular mitigation for future proposed development, and certainly the parties cannot, by an ad hoc agreement, change or exceed an agency's authority. There must be specific statutory authorization or such authorization must be inferred from the more generally articulated statutory authority of the agency. The use of MOUs reflects this uncertainty as to the enforceability of such arrangements, relying upon the expected but non-legally-binding honoring of the terms of agreement. By analogy to the use of "development agreements" and the concept of "vested rights" based upon permits and entitlements previously issued, it can be argued that concepts of equity, such as estoppel, would support the position that regulatory agencies should be and are bound by commitments set forth in MOUs. Further, if an agency can be bound based upon such equitable concepts, it should be able to provide more than "cold comfort" and to enter into agreements that reasonably promote the statutory purposes of the agency. At the same time, it cannot be denied that specific state-level authorization for development agreements in California, Florida, and other states has increased their use and confidence in their effect.

At the other extreme, the Corps once suggested, under the prior Mitigation Banking Regulatory Guidance Letter (RGL), that the credit purchasers could be liable for a bank's failure. Such an approach would inhibit developers from purchasing credits from a mitigation bank for fear of being held liable for the errors or omissions of the bank sponsor over whom the end-user has no control. Although it is true that the purchaser could require indemnification from the banker, this is likely to be a poor remedy if the developer's project is at stake or the banker is insolvent. The only reasonable approach is to limit the liability for bank failures to the bank sponsor.

The regulatory agency's primary interest in enforcing a bank MOA is to remediate the damages to the conserved habitat or wetlands that otherwise would have been provided. Absent any enforcement provisions in the MOA that allow the agency to take affirmative action, or a link between the MOA and a government permit, the regulator may be limited to contract remedies against the mitigation bank sponsor, that is, money damages. These money damages could be utilized to remediate a failed bank. Potential difficulties associated with contract enforcement actions include the fact that the government would be unable to impose sanctions against the mitigation banker and that, under contract law, specific performance and punitive damages are generally not available.[23] Accordingly, some form of enforcement mechanism that allows the agency to take affirmative action should be included in the MOA and specified in detail.[24] For example, the MOA might provide for the conveyance of a conservation easement to the state wildlife agency that expressly provides to that agency the right, in the event of a default in the maintenance obligations of the bank operator, to enter the bank lands and complete the conservation work and thereafter manage the bank lands. The right of the agency to perform such remediation would be conditioned upon, for example, notice to the bank operator of the default and provision of an opportunity to cure.

Notes

1. See, e.g., 30 C.F.R. §§ 330 et seq. (1993).
2. Environmental Law Institute, *Wetland Mitigation Banking* 47 (1993). Citing Cathleen Short, Mitigation Banking (United States Fish and Wildlife Service Research and Development, 1988).
3. Id.
4. 33 U.S.C.A. § 1344(e)(2) (1986).
5. See Federal Guidance for the Establishment, Use and Operation of Mitigation Banks, 60 Fed. Reg. 58609 (1995).
6. Environmental Law Institute, *supra* note 3, p. 48.
7. O.R.S. §§ 196.600-196.665.
8. Environmental Law Institute, *supra* note 3, p. 49.
9. See, e.g., 16 U.S.C.A. §§ 1531 et seq.
10. San Bruno Mountain Habitat Conservation Plan, Endangered Species Act Section 10(a) Permit #PRT 2-9818.
11. Id.
12. 60 Fed. Reg. p. 58609
13. See CF–GC §§ 1775–1793 (1991).
14. 60 Fed. Reg. p. 58608.
15. Id.
16. Id., p. 58609.
17. Id.
18. Id.
19. For a discussion of SAMPs, see chapter 8.
20. CGC §§ 50060 et seq. (Supp. 1994). In any case these assessments are not to exceed $25 per parcel per year adjusted for inflation.
21. See, e.g., California Environmental Quality Act of 1970, Cal. Pub. Resources Code §§ 21000 et seq. (1983).
22. "No Surprises: Assuring Certainty for Private Landowners in Endangered Species Habitat Conservation Planning," Department of Interior Policy, August 11, 1994.

23. Regulators could impose a form of punitive damages against a mitigation banker in default by including an enforcement mechanism in the bank MOA that would permit the regulators to reduce or eliminate completely the number of wetland credit available for sale from the mitigation bank. This mechanism, however, will not permit the regulators to remediate the mitigation to recover the functional values of the wetland credits already sold.

24. The Army Corps of Engineers Institute for Water Resources has suggested that provision for the ability of private parties to bring citizen suits to enforce a bank MOA may be appropriate. *Wetland Mitigation Banking*, p. 91. However, as the institute readily admits, exposure to citizen's suits would most likely be seen as too great a risk of liability for a private entrepreneurial mitigation banker to accept. Id., p. 90 n. 78.

8

Wetland Mitigation Banking and Watershed Planning[1]

John W. Rogers

Introduction

As the limits of a social philosophy rooted in subduing the Earth and making economic use of all resources have become evident, the concept of sustainable development has been gaining widespread acceptance as a vision of how we should function in our society. Although disagreements exist on the sustainability of particular resource and land use practices, a consensus is emerging on the central objective of sustainable development: economic activity that leaves an undiminished and unimpaired stock of environmental goods—rich topsoil, clean air, potable water, healthy forests, and diverse plant and animal species—to future generations. Resource-depleting and environmentally destructive practices need to be replaced with strategies designed to restore resources, establish sustainable rates of use, and foster stewardship of the landscape. Watershed planning, wetland restoration, and mitigation banking are key elements in the process of finding an effective balance that provides environmental quality and economic vitality.

Perhaps the most significant environmental impacts stemming from European colonization of North America have been the removal and fragmentation of native vegetation, which has altered many of the processes necessary for the continued survival of working ecological systems. In many watersheds, water and nutrient cycles are no longer in balance. Natural functions that help protect against flooding and soil erosion, as well as filter and adsorb nonpoint pollution, no longer work effectively. For example, approximately 85 percent of the wetlands that existed in the Mississippi Valley before the arrival of Europeans have been destroyed. Most were drained to grow crops. The estimated \$12–\$16 billion in flood damage caused by the great flood in 1993 could have been held to a minimum if those wetlands were still in place.

The trend continues. The movement of the American population from urban to suburban and exurban areas continues unabated, gobbling up forests, farmland, and wetlands and fragmenting the landscape. This is evident in the Chesapeake Bay watershed, where population density has fallen over the last 20 years, but the average per capita consumption of land has actually increased, as people spread out over the landscape. According to the 1990 U.S. Census, the average land consumption doubled over the last 20 years in the Chesapeake Bay watershed, while population density declined during this period from .65 persons to .33 persons per acre.

Wetlands that remain today are often residuals from development process; many are functionally degraded. Thus, the mix of wetland areas and types that exist in a watershed today may not be the mix that best serves watershed restoration goals, especially in the face of anticipated development pressures. Identification of these conditions in the design of programs to restore and manage wetlands is one purpose of watershed planning. Such planning could integrate regulatory and nonregulatory wetland rehabilitation and protection programs toward the goal of restoring an entire watershed. Mitigation banking could be an integral part of watershed plans, facilitating ecological restoration and economic development.

Restoration often involves replacing some of the vital components of ecosystems that have been lost due to fragmentation. The two primary means for helping restore ecological processes to such fragmented areas is to develop a series of biological nodes large enough to sustain breeding populations of wildlife and corridors that link these areas in form and function. The corridors frequently follow streams and drainageways but may cross upland areas as well. The successional nature of wetland systems make locating mitigation sites within this interconnected watershed network critical.

Wetland mitigation banking sites can become their own biological node, or help expand existing nodes or corridors, as the ecological system is reestablished and reconnected. Wetland mitigation banking can play an important role in helping to restore the "ecological infrastructure" by expanding or creating nodes and filling "gaps" in corridors to maintain ecological systems and functions within watersheds. A watershed plan will help avoid the practice of creating small, isolated wetlands that are surrounded by development and cut off from a watershed.

The mirror image of this ecological model is the system of the built environment: the towns (nodes) and transportation routes (corridors) of human culture that are vital to our social and economic success. Within the framework of sustainable development, nodes and corridors for ecosystems and water quality are highly valued like towns and roads are for human settlement and economic growth.

Mitigation banking can also be part of a dynamic watershed strategy where both public and private land can be used to meet resource management objectives. The manner in which such banks are established and operated may change over time to accommodate the needs of public and private landowners. For example, a mitigation bank may allow companies to selectively cut timber to keep jobs available during certain periods.

There are two ways in which watershed planning could facilitate private commercial mitigation banking. First, watershed planning could reduce the likelihood of restoration project failure. If a plan identifies the presence of conditions that affect mitigation bank sites, then the placement and design of such sites would be improved. If banks sited according to watershed plans are less likely to fail, then regulators might be able to reduce the amount of financial assurance required. This would induce more private investments in these banking efforts. The market for these credits need to be set at a level that ensures the banks will in fact be funded at a level that will allow restoration. In this case, low bids do not count, what counts is the replacement of beneficial functions.

Second, the existence of watershed plans could pave the way for adding flexibility in the regulatory program through the development of wetland categorization systems. For example, the mitigation sequencing rules could be relaxed for certain wetland types in certain locations. In general, one category of wetland would be those of exceptionally high ecological value to the watershed, with functions that are costly and difficult to replace. Such wetlands would be identified in watershed plans. For those areas, avoidance is the best management strategy and only the most water-dependent and high-value development would be even considered for a permit. Another category of wetlands would be degraded wetland sites where cost-effective restoration of functions is possible. Such wetlands could be earmarked as mitigation bank sites.

A greater level of flexibility in applying the mitigation sequencing rules might be warranted at certain sites. This may also include density bonuses for developments that allow for mass transportation and the like. Such areas could be identified in watershed plans. In this manner, bank entrepreneurs would be able to relate their assessment of development demand to the wetlands in their areas and judge the regional demand for mitigation credits.

Also, from the perspective of the private credit suppliers, the current mitigation sequencing rules—which seek to direct development away from all wetlands and which emphasize securing on-site and in-kind mitigation for wetlands losses—will limit the number of permits issued and lower the demand for permits and credits. Conversely, if watershed planning processes facilitate off-site and out-of-kind mitigation for certain wetland categories, this would encourage private commercial banking as a means to meet regulatory goals of "no net loss" of wetlands.

Since Darwin's study of island ecosystems in 1835, biologists have been aware, that a relationship exists between habitat size and species composition. Today's forests, wetlands, and other sensitive ecosystems often exist as biological islands in a sea of disturbed land. Even the largest national parks are not immune to changes in the surrounding landscape. Habitats that are too small lose species at accelerated rates due to inbreeding and surrounding disturbances. In general, an area that is 1,000 acres in size is many times more valuable in protecting and enhancing species diversity than ten 100-acre parcels. If forests are too small for breeding populations, some scientists predict that many species will last less than 50 years.[2] For example, in the last 20 years, many songbirds no longer migrate as far north as Virginia, Maryland, and Pennsylvania in the spring, due to the clearing of forests in northern South America and along the East Coast.[3]

Species extinction is a part of nature. Scientists understand that extinction occurs because of sustained disturbance. What most people do not realize, however, is that during the last 150 years, the rate of alteration of the landscape and ecological systems has dramatically increased, resulting in an unprecedented loss of species and degradation of natural communities.

The eminent ecologist, E. O. Wilson estimates current global rates of extinction at two species per hour. As many as half the species on Earth could be lost over the next few decades, a rate between 1,000 and 10,000 times faster than ever witnessed, making this mass extinction event one of the largest in Earth's history (Wilson, 1988). Although the plight of the individual species has been the focus of public interest, the health of the larger interconnected community of plants, animals, and microbes as ecosystems is perhaps the best gauge of ecosystem vitality.

In 1992, the federal government initiated a review, called the National Biological Survey, of the health of the American landscape. The results of the survey will be published in 1995. Preliminary results indicate that scores of ecosystems of varying types and sizes have declined on a grand but largely unrecognized scale. As the remnants of ecosystems vanish, species adapted to them probably will vanish as well. According to the authors, "Our results indicate that more biodiversity at the ecosystem level has been lost than is generally recognized in environmental policy debates." Despite "uncertainties and unevenness of data" the authors conclude that the "information portrays a striking picture of endangerment." Thirty of 126 ecosystems identified in the study have declined over 98 percent of their area and are considered "critically endangered." Most of these imperiled areas are in the eastern United States.[4]

When managing natural systems, ecosystem management objectives must be balanced with social values and goals. Yet, our knowledge of ecosystems is evolving. Therefore, the important resource management questions of our day can not be addressed with absolute answers. Rather, they must be addressed based on the best current knowledge and values, with full knowledge that the policies and strategies adopted are never final.

Processes in Watershed and Ecosystem Management

Watershed management, ecosystem management, integrated resource planning (IRP), and biodiversity planning are allied concepts. Arguably, these phrases describe concepts or programs that have extensive overlap in geographic areas of concern, philosophical approach, goals, characteristics, and technical strategies. Table 8.1 compares these four basic approaches to water and ecological planning. Mitigation and restoration strategies can play a role in all four processes.

Watershed management takes a holistic approach to improving and maintaining the quality of a water basin. The aim of watershed managers is to reduce the need for filtration and chemical treatment in preparing water for human consumption and habitat needs. Today, water supply watersheds are viewed as assets that can be managed to reduce or eliminate the need for treatment, which itself can generate undesirable treatment by-products. This approach, while not new, is receiving renewed attention. The Environmental Protection Agency (EPA) is coordinating and funding local watershed initiatives. Watershed management can be

Table 8.1 Comparison of Ecological Management and Planning Methods[a]

Goals, Characteristics, or Strategies	Watershed Management	IRP	Ecosystem Management	Biodiversity Planning
Minimizing adverse impacts in the landscape	X	X	X	X
Understanding natural habitat and physical relationships	X	U	X	X
Having a management strategy for achieving goals that combines scientific and policy planning decisions	X	X	X	X
Recognizing that uncertainty and change are endemic to long-range planning	X	X	X	X
Opening the decision-making process to new ideas and interests dealing with limited resources	X	X	X	X
Evaluating planning decisions from a regional and societal perspective as well as an agency perspective	X	U	X	U
Establishing multiobjective goals that include ecological, community, and business needs	X	U	X	U
Consensus process	X	X	X	X
Looking at multiple options for efficiently matching resource supply to consumer demand	X	X	U	
Least cost planning	U	X		
Restricted to a specific watershed	X			
Focused on a specific ecosystem			X	

[a]X, always includes this characteristic; U, usually includes this characteristic.

seen as ecosystem management, biodiversity planning, and integrated resource planning within a watershed.

Integrated resource planning (IRP) is very much like watershed planning except it is not geographically defined. IRP's principal concepts, demonstrated in and borrowed from the electric utility industry, have been expanded and adapted to natural resource parameters and characteristics. IRP is a consensus decision process that deals with limited resource issues. The central feature of IRP in water resources projects is to try to satisfy the needs of a variety of stakeholders with competing interests. The supply-demand aspect of IRP does not dictate that resources are consumed in the traditional sense of the word. Rather, supply is what is wanted by those making the demands. Wilderness, for instance, can be a "supplied" resource where the public is the consumer. Two good examples of IRP programs are the Santa Clara River Study in southern California and the Central Utah Water Supply Study.

The concept of integrated resource planning deals with conservation as well as reuse of water. Wetlands provide treatment of many nutrients and chemicals and can help provide for increased water quality and supply. Mitigation banking can be used in conjunction with water reuse and supply planning on a subwatershed or watershed basis.

In its ideal form, ecosystem management is the human element of altering, by careful manipulation, the communities of all living organisms and all of the physical and biological components. Together, these biological and physical components make up the environment within a previously specified boundary, which provide the long-term sustainable balance that allows for the maximum benefit of historically native species and environmental factors influencing those species, with preference given to none. Ecosystems may be as small as a pond or as large as an entire watershed.

Biodiversity planning deals with habitat conservation to sustain a wide variety of ecosystems and species that may or may not be near extinction. This concept is similar to ecosystem management but is not limited to one ecosystem or to water resources issues.

Watershed and ecosystem management programs have been implemented in many parts of the country. Two well-established programs are the New Jersey Pinelands and the Chesapeake Bay. These are among the most informative models because of their relatively long histories. The Pinelands and Chesapeake Bay experiences demonstrate the value of some guiding principles that any program would be well advised to adopt. These are:

- Clearly expressed and understood needs
- Clearly expressed goals

- Good research
- An open climate for discussion of issues
- Genuine partnerships among stakeholders

In 1988, a panel of experts from the development and scientific communities, government agencies, and universities joined together to develop a vision for the year 2020 for the Chesapeake Bay. The vision is clearly and simply stated. It is presented in the present tense to emphasize what could happen if appropriate actions are undertaken today.

Chesapeake Bay Vision for the Year 2020

Well before the year 2020, the State Comprehensive Development and Infrastructure Plan has been developed and implemented. State and federal agencies, counties, and municipalities encourage diverse and effective land development patterns—ones that concentrate development in urban, suburban, and already developed rural centers. All growing areas have existing or planned facilities. Densities in most of these areas support mass transportation, van pooling, or other forms of ride-sharing to reduce traffic.

These thriving urban centers and suburban areas are supplemented with funding adequate to maintain or enhance existing services. Cities and towns are vitalized by prudent public and private investments. Developers are offered incentives to provide greater community services and mitigate environmental impacts.

New mixed-use growth centers are planned to take advantage of existing and projected infrastructure. Large open-space areas are located within walking, bicycling, or short driving distances for most people. Open space amenities are given the same priority as infrastructure.

Sensitive areas are protected from encroachment and damage. These areas have been defined and mapped by state and local authorities, and effective programs are in place to protect and manage these assets. Very sensitive areas are in public ownership or under easement. Wetlands and lakes, rivers and other water bodies are protected from upland impacts by undisturbed vegetated buffers. In both urban and rural areas the shoreline of the bay and its tributaries form a series of vegetated corridors. These connect to large forested areas and allow for enhanced water quality, ecological balance, and biological diversity. Water supply has become a statewide issue, and safe, adequate supplies are available from protected groundwater and surface-water sources.

Areas with resource-based industries, such as agriculture, forestry, mining, and seafood harvesting, are protected from encroachment of incompatible land uses. These industries remain important parts of the local and state economy. They have brought their environmental problems under control. Protection of these areas through effective land use controls, reasonable incentives, and innovative funding mechanisms ensures a lasting, diverse economy and resource use options for the future.

Transfer of development rights from one land parcel to another better suited for development is commonplace and is proving to be an effective growth and resource management tool.

Growth in rural areas takes place in existing centers. Rural towns and highway intersections are defined by service boundaries, and development space is provided for an appropriate mix of uses. These centers, with the assistance of state and federal governments, provide adequate sewer and water utilities. Use of on-site wastewater treatment is limited to effectively protect surface water and groundwater from pollution.

Outside these rural centers, residential development is limited so as to retain the economic, ecological, and scenic values of the countryside. Large woodlots and forests are retained and are selectively used for managed forestry, if they are not in the preserves or parks. Quarries and other mining activities occur but are screened from neighboring uses by well-developed wooded buffers. Municipal, county, and state roads are planned to allow for adequate capacity of rural traffic.

The volumes of waste produced in the region have been greatly reduced and are being effectively handled. Energy and water use per capita has been reduced as conservation programs have been put in place. The public and government agencies are sensitive to their responsibilities not to damage the environment and to conserve resources.

Stewardship of the land and bay is practiced by ordinary citizens who have been made aware of how they affect the land and water. The quality of the bay is improved; tourism is strong; and resource-based industry, manufacturing, and service businesses desire to locate in the basin because of its resource base, amenities, diverse economy, and the quality of life it provides to residents.

Programs are supported by the Development and Conservation Trust Fund which funds infrastructure and purchases land, easements, and development rights in support of the goals of the State Comprehensive Development and Infrastructure Plan.

The Chesapeake Bay vision provided a framework to make useful recommendations. The panel developed six linked principles that are reinforced by the vision of what should come to pass in the region by the year 2020:

- Development is concentrated in suitable areas;
- Sensitive areas are protected (managed at appropriate levels);
- Growth is directed to existing population centers in rural areas and resource areas are protected (managed at appropriate levels);
- Stewardship of the bay and the land is a universal ethic;
- Conservation of resources, including the reduction of resource consumption, is practiced throughout the region;
- Funding mechanisms are in place to achieve all other principles.

All segments of society would benefit from the achievement of these principles. Likewise, all must share in the cost of their implementation.

The principles of the Chesapeake vision are evident in other areas of the country. For example, Florida has adopted a policy of moving toward ecosystem management in its planning, land acquisition, and regulatory arenas. Its working definition of ecosystem management is "an integrated, flexible, approach to management of Florida's biological and physical environments—conducted through the use of tools such as planning, land acquisition, environmental education, regulation, and pollution prevention—designed to maintain, protect, and improve the state's natural, managed, and human communities." In many cases, watersheds are the appropriate unit for ecosystem management. The idea of a Development and Conservation Trust Fund for the Chesapeake Bay is similar to Florida's Preservation 2000 plan where the state would raise money to place resource protection at the same level as infrastructure development.

Determining the Relationship, Size, and Function of Mitigation Banks

Widely diversified ecosystems, those with a great variety of plant and animal species, are much more likely to survive periods of environmental stress than are ecosystems with little diversification. Small, isolated areas are more vulnerable than large ones to extinctions from disease, interbreeding, and chronic habitat disturbance from human and natural causes. When wooded areas are broken into isolated pieces, it is difficult for organisms to move from one habitat fragment to another.

Studies have shown that on average, a habitat will lose between 30 and 50 percent of its species for each 90 percent reduction in area. For example, a 100-acre "island" will retain only about 70 percent of the species that were able to survive in a 1,000-acre habitat.[5] Research and population dynamics provide the basis for patterns and minimum forest sizes to support various levels of diversity.

Figure 8.1 depicts ecological land-planning principles aimed at maintaining a pattern of large forests and corridors for the purpose of sustaining biological diversity. The drawings represent natural islands of forest and wetland and streams among developed towns and farmlands. These six pairs of drawings emphasize the ideal (left column and top) compared to the less preferred (right column and bottom). More recently, these associations also have been documented in habitat types other than upland forests, specifically wetland habitats.[6] Scientific research supports the important connection between wildlife management and land use planning.[7]

Areas of less than 35 acres have significantly fewer species, while areas of over 175 acres have significantly more species.[8] Ruptures usually require several areas of 600 to 1,000 acres.[9]

The variety of habitats that woodlands provide depends on more than the size of an area. Depth is important, as is the extent of natural boundaries or edges between ecosystems. A wooded area must be at least 100

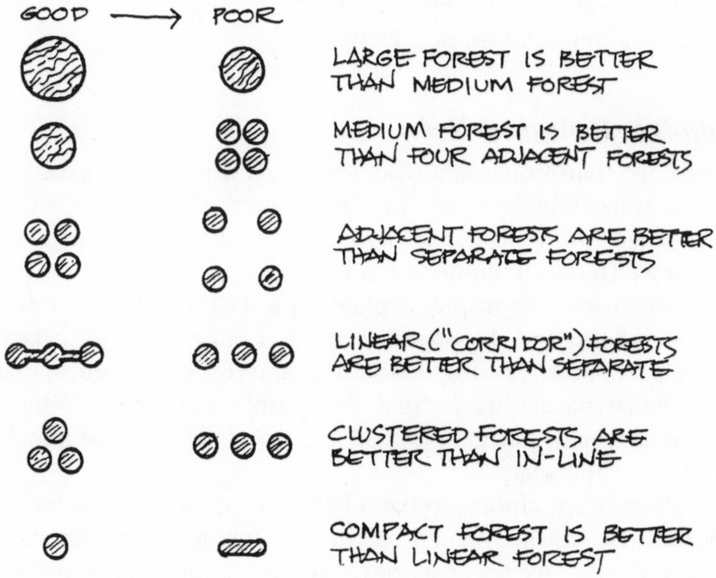

Figure 8.1 Habitat Island Design Principles

meters wide (about 300 feet) to provide a deep-forest habitat. Edges between forests and fields are ecologically diverse areas. A wide variety of animals is found there: those that come from each of the abutting habitats, those that make frequent use of more than one vegetation type, and those that have adapted to life on the edge habitat itself. This is called the "edge effect."

Restoring wetlands as buffers to existing wildlife areas can mitigate the edge effect, expand the effective size, improve the likelihood of self-regulation, augment biological diversity, and safeguard these areas from surrounding disturbance.

Using mitigation banks to serve as corridors between existing natural areas expands habitat for many larger species who need wider ranges to survive, and can increase species mobility between natural islands to avoid isolating populations. The New Jersey Pinelands Plan used this concept of nodes and corridors for the basis of its comprehensive resource management plan.

The amount of edge in a given wooded area varies greatly with its shape. A circular area has the least edge. Elongated areas or areas with indentations and protrusions have more edge. Unfortunately, the "edge effect" that benefits game animals often harms nongame species. Although some common nongame animals such as raccoons, mockingbirds, and opossums may prosper with increasing local habitat variety, many rare and sensitive species disappear.

Wetlands as Biological Nodes

Wetlands are transitional lands between terrestrial and aquatic systems where the water table is at or near the surface or the land is covered by shallow water. Although regulatory agencies have slightly differing definitions of wetlands, all require the presence of one or more of three wetland characteristics: hydrophytic plants (plants that grow in wet soils), hydric (wet) soils, and the presence of water at or near the surface during some part of the growing season. Wetlands include swamps, marshes, or bogs and typically are located along stream corridors, ponds, and lakes. Wetlands also occur in upland and forested areas and may be dry during part of the year.

The relationships among wetland habitats are similar to those among upland habitats: smaller habitat patches support fewer species and more isolated wetlands have fewer species.[10] Although these relationships have been studied most intensively for birds, the number of mammal species also increases with habitat size.[11] Studies found significantly more species of mammals in areas over 100 acres than in smaller areas. Large mammals typically require large contiguous areas, 1,000 acres and more.[12]

Table 8.2 Site Classification for Protection of Biological Diversity and Ecosystem Function

Class	Description	Defining Objective
Class 1	Globally significant elements	Preservation and recovery • Occupied habitat of significant communities • Sites critical to life cycle • Suitable habitat likely to be occupied • Unique example of community types
Class 2	Regionally significant elements	No decline or degradation • Occupied habitat or significant communities • Sites critical to life cycle • Suitable habitat likely to be occupied
Class 3	Greater that 500 ha (acres), supporting naturally occurring species assemblages	Maintain native species composition and ecological process • Increased flexibility • Landscape context • Management at community level • Minimize edge • Ecological restoration option
Class 4	Less than 500 ha (acres), supporting some natural communities	Maintain native species composition and ecological process • Smaller size and/or lack of connectivity • Minimized edge • Stewardship at coarser level • Elements in danger of local extirpation
Class 5	Isolated, less than 125 ha, supporting no elements of special significance	Maintain Native Species Most commonly encountered site class important local functions • Open space • Recreation • Model for public relations and education Enhancement for Native Wildlife • Nest boxes • Food plots • Feeding stations
Class 6	Isolated degraded parcels with few native species and no natural community	No further degradation • Avid loss of native species • Restoration avoiding further damage • Target predisturbance community

Source: World Wildlife Fund, Managing Private Lands to Conserve Biological Diversity, 1991.

Current patterns and sizes of local woodlands, wildlife management areas, and park networks often may be inadequate to protect many species beyond the short term. The present regulatory approach to protect plant and animal species is toward "endangered species" rather than ecosystems as a whole. Although they deserve protection, particular endangered species may not be as important at the local level as maintaining an ecosystem where the whole range of local native plant and animal life can exist and flourish. The World Wildlife Fund has developed several classes of land (see Table 8.2) in their evaluation of areas to be protected.[13]

Mitigation Banking in Watershed Planning

In the United States over two-thirds of the land is privately owned. Wetland mitigation banking, in the context of watershed planning and ecosystem management, provides a mechanism to harness the energy, creativity, and resources of the private sector in designing integrated land management approaches that build on existing public ownership land use patterns.

One of the obstacles to creating successful mitigation banks is the inability of agencies to consider adequately the long-term ecological significance of a particular site. Too often regulatory agencies have been caught in the project-by-project permitting mode without enough emphasis on broad ecosystem issues. This is especially problematic where watersheds contain many jurisdictions, each with a separate planning responsibility. To ensure sustainable ecological functions, watershed planners need to take an aggressive approach that involves all stakeholders in the planning process and sets priorities for locating restoration sites and mitigation banks.

The philosophical approach to wetland mitigation and watershed planning can vary greatly. Table 8.3 summarizes approaches that provide a framework for wetland mitigation banking. Each approach has specific advantages and disadvantages. Watershed planning and wetland banking must show consistency or alignment in purpose. To establish any type of prioritization process for ranking or selecting mitigation sites, there must be a basic approach in mind.

Mitigation banking can serve a variety of ecological and development goals. To a wetlands ecologist, bank siting should maximize ecological functions, for example, by replacing the most disturbed wetlands near the headwaters of a river or stream. To private developers, however, mitigation banks should be created on a wide range of sites to maximize the supply and minimize the price of credits.

Table 8.3 Approaches to Watershed Planning for Wetland Mitigation Sites

Approach	Advantages	Disadvantages
Prehistoric baseline. Assumes knowledge of historic conditions and meaningful mitigation. EXAMPLES: West Eugene Oregon Preservation Plan; Public Service Electric and Gas (PSE&G) Wetlands Mitigation project in the Delaware River.	Provides opportunity for large off-site, out-of-kind mitigation. Differential compensation ratios and service area requirements can assure that wetland losses are compensated for in a way that makes sense in the regional landscape. This approach provides for greater likelihood of success in creating the same type of wetland where it once existed. Other unrecognized values may be found by restoring original vegetation patterns.	Difficult to justify. Nature is constantly changing. Ecosystem management is more about managing the rate of change than holding things constant. Implementation problems come from a lack of data to guide restoration and lack of plant material. Demand may be inconsistent with necessary types and locations of compensatory wetlands.
Maximize an array of functions. Involves organizing wetland banks to produce a variety of wetland functions. EXAMPLE: Batiquitos Lagoon, Port of Los Angeles.	This approach is useful where a variety of small wetland types need to be replaced. This approach should provide the most wetland benefit to the watershed.	Learning about how to replace the many functions can be difficult and very costly.
Maximize a single function. Generally easier and less expensive to accomplish than trying to maximize and balance many different functions. EXAMPLE: Florida Regional Wildlife and Wetlands Conservation and Mitigation Area in Florida.	Monitoring should also be easier and less costly; determining success is simpler as well. This approach may be appropriate where the identified need is for flood control, or wildlife corridors, or sediment trapping, or a particular wetland type.	Wasting potential opportunities to replace other functions. Biasing goals toward easy solutions rather than ecological goals.

Source: Wetland Mitigation Banking IWR Report 94-WMB-6.

Mitigation banks are attractive chiefly because they offer an opportunity to make compensatory mitigation more ecologically significant and convenient. Moreover, mitigation banking helps create incentives for restoring wetlands. Even if a banking system is driven by private decisions of mitigation credit producers and clients, however, the framework under which they operate should be designed to assure that ecological values are captured in siting decisions.

If created wetlands are replacing natural wetlands, the regulatory agencies are not succeeding in their goal of "no net loss," given the poor track record of created wetlands. Developers are to blame for many failed mitigation efforts. The present regulatory policy, which encourages mitigation on-site, has created a "lose-lose" situation for both the environmental community and developers. The widespread failure of on-site, project-by-project mitigation suggests that the agencies develop a mitigation policy that encourages the use of large regional wetland mitigation banks. The potential benefits are substantial, yet several concerns must be addressed.

Opponents of mitigation banking suggest that shifting the focus of replacing lost wetlands function on-site to a mitigation site elsewhere destroys the viability of a functioning wetland in an area where its resource value may be of greatest benefit. However, such criticism ignores the underlying wetland protection and mitigation sequencing regulations. That is, because avoidance is now the primary criteria to be used as part of an alternatives analysis review, disturbance of the wetlands can occur only after the avoidance alternative has been appropriately considered and rejected.

Of the nine states that have authorized mitigation banking by statute, few include any guidelines or reference to siting criteria. Most defer the issue to implementing regulations, which, unfortunately, say relatively little.

At the local level, integrating watershed planning and wetland mitigation banking can occur in the comprehensive planning process. At the federal level, several mechanisms exist to integrate local and state planning into the federal permitting process. For example, when a state or local agency adopts a comprehensive watershed management plan, the U.S. Army Corps of Engineers (Corps) will strive to establish regional general permits or programmatic general permits (PGP), depending on whether a state, regional, or local program exists, that would regulate wetland losses. If there is no other government program that protects wetlands, then the Corps would issue a regional permit based on the plan for activities in wetlands identified as low value. If another government agency program exists upon which the Corps can base a PGP, then the Corps would more likely issue a PGP. In either case, the Corps would focus compensatory mitigation requirements for wetland areas identified in the watershed management plan as priority restoration areas. In the best of circumstances, a mitigation bank could be created by restoring such priority areas. This would not only focus restoration on the priority wetland areas, but also would minimize the regulatory burden on activities authorized by identifying in advance the mitigation through the mitigation bank.

In addition to regional permits and programmatic general permits, the federal government has two other mechanisms at its disposal to integrate watershed planning with mitigation banking: Advanced Identification (ADID) and Special Areas Management Plans (SAMP).

Under ADID, EPA, in conjunction with the Corps of Engineers and after consulting with the state, may identify wetlands as generally suitable or unsuitable for discharge of dredged and fill material, in advance of permit applications. ADIDs are authorized in Section 404(b)(1) of the Clean Water Act and are often funded through EPA grants. EPA selects ADID sites based on the perceived need for advance identification, that is, where EPA feels there is likely to be significant development pressure in areas that contain ecologically valuable wetlands.

The goal of ADID is to make the permitting process more predictable and to provide local officials with some general guidance so they will not approve permits that likely would be denied by the federal government.

As of December 1992, there were 71 ADIDs across the nation, 35 completed and 36 ongoing.[14] The size, scope, and degree of local involvement with these ADIDs vary. While ADID areas sometimes correspond to watershed boundaries, this is not necessarily the case. ADIDs can be initiated by EPA, but they can also be requested by state or local entities in order to facilitate local planning efforts. ADIDs are often components of other plans, such as in the case in West Eugene, Oregon, and Mill Creek, Washington. While EPA emphasizes that ADIDs are strictly advisory, the Corps seems interested in using the ADID process to facilitate its permitting process, when allowable. For instance, following an ADID in Lake County, Illinois, the Corps retracted some nationwide permits that had allowed certain activities in some of the wetlands that the ADID identified as functionally valuable.

ADIDs assess the functional value of wetlands prior to permit applications. An ADID assessment of a site does not grant or deny permits, but it does give some indication of where fill activities are likely to be allowed. In that sense, ADIDs are thought to be useful to developers as they provide advance warning about where permits are more or less likely to be given. ADIDs are useful to regulators as well, as they could expedite the review of individual permits by providing regulators with a database of wetland sites and functions. ADIDs are thought to be useful in preventing inadvertent unauthorized filling of wetlands by making landowners more aware of wetlands on their property.

Advance identification of wetlands could also contribute to private mitigation banking, helping bankers assess the likely demand for credits and identify appropriate mitigation sites. However, in some cases, ADID projects have experienced problems. The advanced identification process itself sometimes proves difficult, due to scientific uncertainty or the sheer geographic area of some ADID sites. Moreover, different interests sometimes voice opposition to a given ADID. Although advanced categorizations are not binding, in some instances landowners believe that advanced identification of sites unsuitable for fill reduces the value of their property. On the other hand, environmentalists and some regulators occasionally oppose advanced identification of wetland sites as suitable for development because they believe that the designation encourages development and reduces protection of these wetlands.

SAMPs, established under the 1980 amendments to the Coastal Zone Management Act (CZMA), are "comprehensive plans providing for natural resource protection and reasonable coastal-dependent economic growth." Like ADIDs, SAMPs may or may not correspond to watershed boundaries. However, SAMPs are more comprehensive than ADIDs and emphasize multiagency and public participation. Also, unlike ADIDs, approved SAMPs have formal legal status and may serve as the basis for permitting decisions. Although SAMPs apply only to the coastal zone, the Corps has applied the SAMP procedure in inland areas. The Corps feels it has the authority to do this based on Section 404 of the Clean Water Act, which gives it authority to grant general permits for certain activities. In general, the Corps participates in the development of SAMPs when there is (1) significant development pressure in environmentally sensitive areas, (2) local involvement, (3) a participating local agency, and (4) an agreement of all parties on the outcome of the plan. It appears that this fourth point has proven the most difficult to obtain; often there is disagreement among agencies and among property owners, commercial interests, and environmental groups.

SAMPs are potentially useful to mitigation banking in ways similar to ADIDs. SAMPs could categorize wetlands. Once accepted, however, categorizations would be binding. This would add certainty to any mitigation banking elements of plans (if one is included) if a wetland category specifies that mitigation can be met through banking. For example, the West Eugene Wetlands Plan, described by the Corps as a SAMP, is expected to establish wetland categories that specify those areas that will receive permits if they purchase credits from a (public) mitigation bank.

Delegating the Corps's authority to the local government is expected to expedite the permitting process, but it also raises questions about the integrity and tenacity of the permitting decisions. The implementation of a comprehensive plan is only as good as the quality of ideas and data in the plan. This also raises questions about the ability of local governments to apply the rules consistently. Planners must have clear and defensible goals and criteria, and the Army Corps of Engineers must work to ensure that it assesses permit applications consistently.

Criteria for Siting Mitigation Banks

The Homebuilders Association of Greater Chicago has proposed detailed siting criteria for a mitigation bank.

- Sites containing highly disturbed lands or prior converted wetlands, with a preference given to sites with high restoration potential
- Sites with no high-quality wetlands present where wetlands could be created
- Sites containing some upland areas to provide diverse habitat
- Sites where adequate hydrology can be secured
- Sites near or adjacent to large public landholding to increase effective bank size

Probably the most detailed site selection work has been conducted by Public Service Electric and Gas (PSE&G) Company of New Jersey, which is in the process of restoring 10,000 acres of wetlands (basically Spartina marsh) along the Delaware River in lieu of a NPDES (National Pollution Discharge Elimination System) permit for relicensing a power plant. The bank is unique in that it is not being created to replace wetland losses but to compensate for the fish that are killed during the cooling process for PSE&G's power plants. The state of New Jersey has decided that the creation of additional fish habitat through wetland restoration will offset the fish losses due to operation of the power plant. The primary site selection criteria follow very closely with those just listed. Only disturbed wetlands will be restored.

Where ecological enhancement is the goal, the following site considerations are generally considered valuable:

- Bigger is better—larger areas provide a greater variety of vegetation types and successional stages and thus greater potential for maintaining biological diversity. They also provide for more management opportunities and sanctuary for animals at different times in their life cycles. Highly disturbed sites surrounded by fully functional wetlands or other large holdings, such as wildlife sanctuaries or parklands, are most desirable. Based upon the World Wildlife Fund Site Classification for managing biodiversity on private

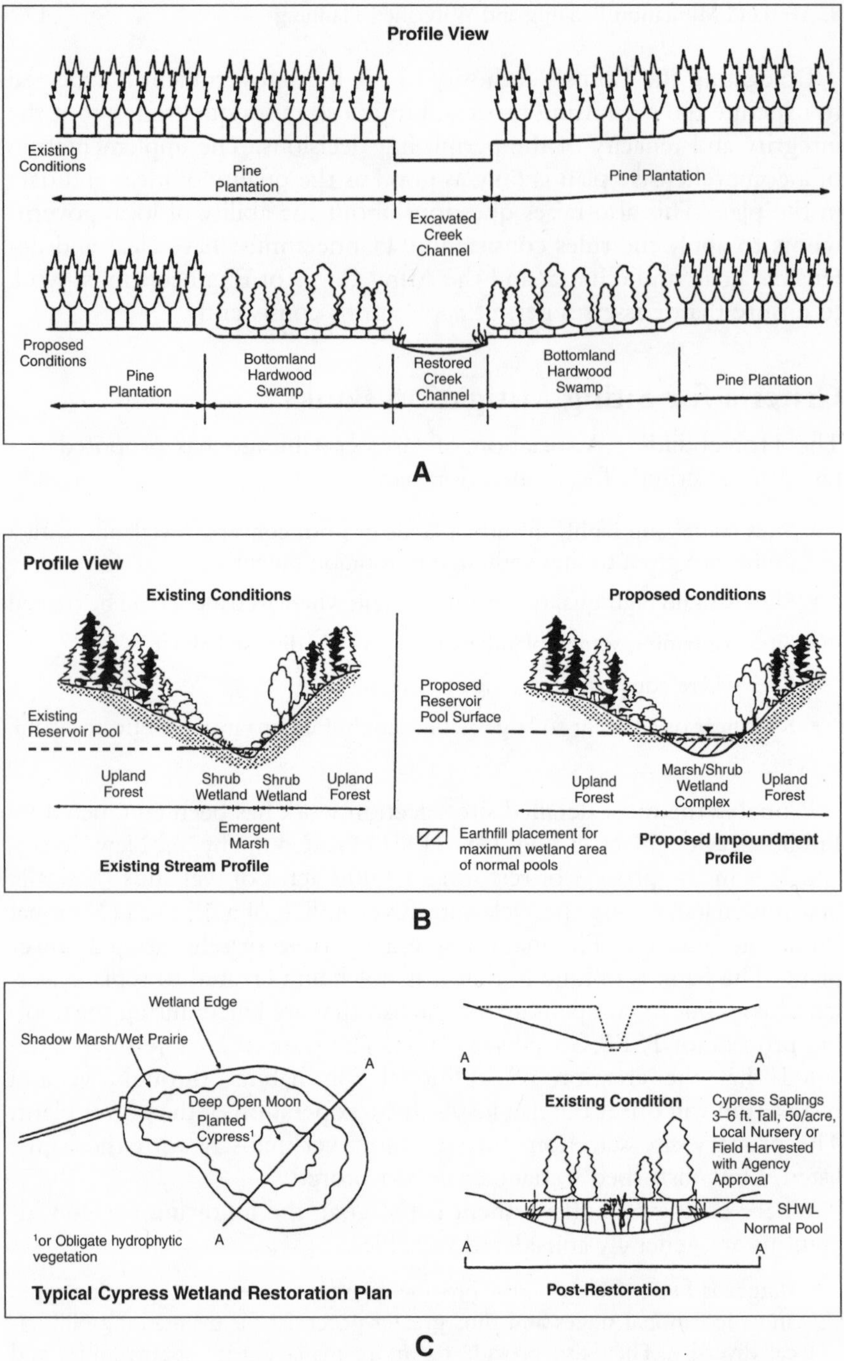

Figure 8.2 Conceptual Plan Profile for (A) New Reservoir Mitigation Bank, (B) Reservoir Expansion Mitigation Bank, and (C) New Well Field Mitigation Bank

lands, an isolated area less than 125 acres in size, with no significant features, would have limited value as a biological island.

- Unique is valuable—unique natural areas that are substantially different from surrounding areas in terms of vegetation, water availability, or terrain can add greatly to the diversity on site. Species-rich transition areas, such as streams and river banks, also are very valuable. Some small wetland or habitat areas, such as vernal pools and prairie potholes, are very valuable in maintaining biological diversity.
- Protect the sensitive—protecting threatened or endangered species that are sensitive to specific types of disturbance is very important.
- Native species are preferable to exotic—diversity should only be measured in terms of native species, endemic to an area. Selection of sites with non-native species with a goal to totally eradicate and return native vegetation is preferred. Areas covered with common reed grass, *Phragmites communis,* are good locations for wetland mitigation as long as the hydrological conditions are suitable for wetland restoration or enhancement with native plants.
- Maintain variety—maintaining a variety of habitat types and successional stages within a watershed is desirable.
- Maintain buffer areas where possible—maintaining buffers around mitigation sites provide short-term vegetation structure and diversity of habitat. Buffers also provide visual screens from adjacent land uses and can act as barriers to predators and exotic species.

To enhance restoration reliability sites must have

- adequate hydrology to support wetland systems and the hydrology should provide for a self-sustaining system;
- adjacent areas that have seed sources to help replenish the site or adjacent sites that have species that will invade the site causing it to fail;
- salinity in tidal settings.

Other important values might include restoring or creating wetlands to improve flood control. Restoring wetlands in the 100-year floodplain will expand the floodplain's capacity to retain flood waters and will limit flood damage. Many communities already restrict development in floodplains, so acquiring bank sites should be less expensive in those areas. In addition, communities that restore floodplains and/or restrict development in floodplains benefit financially by receiving lower flood insurance rates through the Federal Emergency Management Agency's Community Rating System.

Restoring or creating wetlands along riparian areas, adjacent lakes, ponds, or other waters can serve to filter nonpoint source pollution and improve water quality. The three diagrams in figure 8.2 show how impoundments can use wetland banking principles.

Practical considerations must include:

- Jurisdictional issues—many banking agreements will include multiple agencies or jurisdictions. If these jurisdictions have competing objectives, one wanting improved habitat while others are concerned about property taxes, additional incentives or agreements will need to be developed.
- Site survey—sites should be free of hazardous waste.
- Size—the size will depend on demand for credits.
- Proximity—bank sites should be close to impacted sites where possible. Wetlands perform important functions that are directly connected to their location in the landscape. Therefore, banking sites should be within the same watershed where possible.
- Flexibility—banking regulations should provide enough flexibility to ensure participation. Because the capital cost of acquiring a bank site can be high, banking may only be profitable for private operators when they already own the site. There should be some allowance for ownership in the site selection criteria.
- Consistency—the location of mitigation banks must be consistent with local and state plans.
- Aesthetics—adjacent landowners may view wetlands as a visual impairment or a nuisance that attracts mosquitos. Others may fear the loss of tax dollars. Selecting the best site for a bank involves planning and developing a planning approach and site selection criteria.

Conclusion

Development need not fragment or destroy natural patterns. By maintaining landscape patterns of large woodlands and wetlands and wooded stream valleys and drainage-ways, natural functions can be significantly protected.

To instill effectively a sustainable development ethic will require extensive and continuous education. Developing a meaningful stewardship ethic requires reexamining and reshaping how Americans engage the natural world. This is not work for a few policymakers, biologists, and volunteers; it will require significant debate and agreement among government, the community, and business.

Laws will have to be redrafted to acknowledge the interconnections between resource and social objectives. For example, most laws regarding the natural environment do not mandate protection of biological diversity and are not directed to ecosystems as a whole. Instead, they aim to protect particular resources, such as a species with a dangerously low population level or the quality of water supply sources used for drinking water. Likewise, resource-oriented laws are almost universally silent on economic and social concerns that must be incorporated into resource management.

Private mitigation credit markets could help the federal wetlands regulatory program achieve no net loss of wetlands along with single-user banks. Decisions affecting wetlands are most often made on a project-by-project, permit-by-permit basis. This approach limits the scope of review by federal agencies and limits opportunities an applicant has to use innovative approaches to reconfigure a project.

National guidance should give individual regions the flexibility to produce regional guidelines specific to their own watershed needs, which in turn would increase certainty for mitigation success and prospective credit suppliers in those areas. Of course, regulations in the field must provide information to prospective credit suppliers on the general process and steps required for the regulatory approval of private commercial factors that must be addressed.

Watershed plans should carefully consider the use of biological nodes connected, where possible, to riparian vegetation along streams as a network or "ecological infrastructure." Regulators should use these watershed plans to ensure that the siting of banks and bank design and construction plans contribute to local watershed goals.

The categorization of wetlands to establish functional significance in a watershed should be advanced through watershed planning. Once accomplished, this would enable the regulatory program to relax the mitigation sequencing requirements for those wetlands that have been characterized in watershed plans as suitable mitigation trading.

Sites for mitigation trading should fill the gaps in riparian forests and contribute to the size of biological, multispecies nodes to ensure protection of a wide range of species.

The types and sizes of wetland development projects that may use banks, and the geographic range of the bank service area, should be determined according to area-specific factors of the fill permit.

Performance mitigation standards should be required in order to determine when a bank mitigation parcel is failing or has failed. However, these should account for less-than-extreme natural events, which may cause a bank mitigation parcel to evolve along a somewhat different pathway than originally planned.

Monitoring and maintenance should be required. Bankers should repair and detect deficiencies regarding site construction and performance.

However, the monitoring and maintenance should be limited to a reasonable time frame.

Wetland policy and programs of the regulatory agency should have similar quality control standards for all mitigation projects, whether done on-site or off-site through mitigation banks.

The better the planning and the prioritization of criteria for bank site selection, the more likely the watershed ecosystems and ecological functions will be restored and enhanced. Mitigation banking provides an excellent incentive for private landowners to benefit from restoration and mitigation efforts that are good for the environment and our economy.

Notes

1. The discussion of Advance Identification and Special Area Management Plans was adapted from the chapter contributed by Shabman, Scodari, and King, as well as from Redmond et al.
2. World Wildlife Fund, 1991 "Managing Private Lands to Conserve Biodiversity."
3. Terborgh, Jo, 1992. "Why Songbirds are Vanishing." *Scientific American* 266(5): 98–105.
4. *New York Times,* February 14, 1995. Latest Endangered Species, p. C-3.
5. Diamond, J. M. 1975. "The Island Dilemma: Lessons of Modern Biogeographic Studies for the Design of Nature Preserves." *Biol. Conserv.* 7:129–45.
6. Weller, M. W., and L. H. Fredrickson. 1974. "Avian Ecology of a Managed Glacial Marsh." *Living Bird* 12:269–91; M. Brown and J. J. Dinsmore. 1986. "Implications of Marsh Size and Isolation for Marsh Bird Management." *J. Wildl. Manage.* 50:392–97; J. P. Gibbs and S. M. Melvin. 1989. "An Assessment of Wading Birds and Other Wetlands Avifauna and Their Habitats in Maine." Final Report Maine Department of Inland Fish and Wildlife Bangor, ME.
7. Diamond, 1975; Foresman et al., 1976; Game, M. 1980. "Best Shape for Nature Reserves," *Nature* 287:630–32; A. J. Higgs and M. B. Usher. 1980. "Should Nature Reserves Be Large or Small?" *Nature* 285:568–69.
8. Whitcomb, R.F., C.S. Robbins, J.F. Lunch, B.L. Whitcomb, M.K. Klimkiewicz, and D. Bystrak. 1981. "Effects of Forest Fragmentation on Avifauana of the Eastern Deciduous Forest,"in *Forest Island Dynamics in Man-Dominated Landscapes,* R. L. Burgess and D. M. Sharpe, eds., pp. 123–205. New York: Springer-Verlag, 1981.
9. Fryer-Nurza, Jo-ann, interview, 1990.
10. Weller and Fredrickson, et al., ibid.

11. Brown, J. H. 1971. Mammals on Mountain Tops. "Non Equilibrium Insular Biogeography." *American Naturalist* 105:467–78; J. H. Brown. 1978. "The Theory of Insular Biogeography and the Distribution of Boreal Birds and Mammals." *Great Basin Nat. Mem.* 2:209–27.

12. Burghardt 1976; Rogers and Allen, 1987. "Forest Island Dynamics in Man-Dominated Landscapes," *Ecological Series.* Vol. 41, New York: Springer-Verlag.

13. World Wildlife Fund, 1991. "Managing Private Lands to Conserve Biodiversity."

14. Office of Wetlands, Oceans, and Waterways, U.S. Environmental Protection Agency, "Summary of Advance Identification Projects Under Section 230.80 of the 404(b)(1) Guidelines," Washington, D.C., December 1992.

9

The Practice of Mitigation Banking

Lindell L. Marsh and Jora Young*

Wetland mitigation bankers face many of the same risks as farmers who market produce from their lands, except that mitigation bankers must guarantee the success of their crop in perpetuity. Despite the apparent financial risks, the demand by developers for convenient, readily available mitigation, together with a significant desire on the part of the resource agencies and conservation interests to make the federal Clean Water Act more user-friendly, appears to support a significant and growing interest in mitigation banking. This chapter discusses the practice of mitigation banking, focusing first on the qualifications of the banker and then on the process of establishing and implementing the bank.

Who Should Be in the Mitigation Banking Business?

Unlike farming, mitigation banking focuses on creating, restoring, and/or preserving a wetland or habitat supporting a suite of species

*The authors would like to thank Dr. Terry Huffman for his assistance in preparing this chapter.

often prescribed by a number of local, state, and federal agencies using very subjective criteria, to meet an uncertain demand by the development community for mitigation credits. In short, it is a business based on a complex and varied set of factors, requiring a number of different highly developed skills and capabilities, including the capability to:

- design, engineer, and implement a program that is often of a large-scale and complex nature;
- work with local, state, and federal agencies to obtain government approvals and permits for the site as an established or restored wetland/habitat, while contemplating the regulatory requirements for other development for which the mitigation site may provide credits;
- market the credits produced to developers who require off-site mitigation;
- manage the conserved habitat over the long term.

These general areas of capability require management and marketing expertise but also include the need for more specific expertise, such as legal, accounting, biological, and engineering services.

Because of the variety and high-level expertise required, it is quite common for a mitigation banking enterprise to be undertaken by a team. Further, because such broad expertise is required even for a small bank, the per-unit costs of the enterprise can escalate rapidly, encouraging larger-scale banks or organizations that focus on a number of banks. This complexity also explains in part the historic lack of success of entrepreneurial banks.

For the same reason, single-user banks formed, for example, within a transportation agency have been able to assemble the necessary expertise and efficiently provide for its mitigation needs. This also explains in part the advantage of a mitigation bank formed as part of a large-scale conservation plan and effort, such as the habitat conservation planning effort under the Endangered Species Act.

To the extent that a single individual bank sponsor possesses interdisciplinary expertise, it will normally be more efficient in establishing and operating a bank. For example, a biologist who has assisted developers in successfully preparing mitigation programs satisfactory to the various regulatory agencies within a specific area for a significant period of time would be in a good position to undertake a banking enterprise. The biologist would have a sense of what would be required to make the bank work biologically and economically, be able to work with the regulatory agencies, have an appreciation of the demand for and value of mitigation credits, and have the contacts necessary to market the credits.

Conversely, a businessman from another area with no experience in implementing a biological mitigation program or in working with governmental agencies would be a poor candidate for undertaking a bank.

To date, mitigation banking has been dominated by single-users, such as transportation agencies, pipeline and energy transmission companies, and oil and gas producers. Timber and mining companies are beginning to develop plans that contemplate mitigation banking–like approaches. Entrepreneurial banks are operated by for-profit corporations (see the Florida Wetlandsbank and the Millhaven bank case studies), as well as public sector agencies and nonprofit organizations (see the Coachella Valley case study). Their reasons for being in the mitigation banking business vary. The single-users are satisfying their own needs for mitigation; the for-profit organizations are hoping to reap a return on their investments; and the public agencies are often attempting to enhance publicly owned lands, although, increasingly public agencies are seeing mitigation banking as serving the needs of conservation as well as economic development within their communities. For example, without the Riverside County Habitat Conservation Agency program, it would be extremely difficult for individual developers to efficiently satisfy the mitigation requirements of federal and state agencies, resulting in a severe bottleneck for desired economic growth within the county.

Would-be banks need to assemble a team that will possess the capabilities outlined here. The selection of the team is a critically important decision that would normally precede the acquisition of the mitigation site. Of course, in some cases, a bank may be initiated by the owner of a parcel, whether public or private, who is trying to profit from its ownership. The Carlsbad Highlands bank, for example, was initiated by the Bank of America as a means of recouping its investment from land on which it had foreclosed. Because of the existence of the California gnatcatcher, a federally listed "threatened" species, within a broad expanse of coastal sage scrub habitat on the site, in 1995 Bank of America (BOA) chose to establish a mitigation bank that would preserve the lands, provide mitigation credits to developers within San Diego County and, hopefully, recoup a portion of its investment. The bank will involve preservation of the site: no enhancement or creation is required. Credits may be marketed within a service area comprising San Diego County. Credits are measured on an acreage basis and may be applied to impacts to coastal sage scrub habitat. The number of acres of credits to be required will be determined by the agencies in connection with the respective development approval.

For the long term, the site will be conveyed in three phases to the Department of Fish and Game (DFG) in perpetuity. In addition, BOA will endow DFG in the amount of $1,500 per acre (to obtain an annual yield if invested of 7 percent per annum) for this purpose. Initial credits have been sold to the California Transportation Agency for application to impacts from highway construction. The approval of the project was accompanied by the promulgation of the conservation banking policy of the California Resources Agency and the Environmental Protection Agency.

In cases like this, the assessment focuses on a somewhat different question; that is, whether a mitigation bank would be more profitable (either in monetary terms or in public goals) than alternative uses of the particular lands.

Typically, the banking enterprise performs the following management functions:

- Personnel administration
- Permitting
- Permit compliance
- Credit sales
- Finance
- Land purchase and landowner joint ventures (where the land is not already owned)
- Establishment of conservation easements
- Community relations
- Local, state, and federal resource agency relations
- Contractor selection and management (administrative activities)

Generally, the most cost-effective approach for creating and operating the bank is to keep the permanent bank staff to a minimum and utilize contractors on an as-needed basis to accomplish the operations and maintenance needs of the bank: debit creation (restoration, enhancement, creation, or protection), maintenance, monitoring, compliance reporting, remediation of unsuccessful debit creation, site protective measures, and development of community use and education opportunities. At a minimum, at least one bank employee should have the required technical expertise and hands-on experience in wetlands mitigation. This individual would serve as the manager of the various contractors utilized, as well as serve as the field representative who would interface with resource agency personnel.

The biological and hydrological complexity of the wetlands/wildlife resource involved may vary. A site having relatively simple biological conditions may require the attention of biologists who are knowledgeable about specific species or a suite of species. More complex conditions may require the engagement of a biologist with a broader view of watersheds and ecosystems and the multiplicity of species involved. The qualifications of the supervising biologist must then include the ability to design and manage a scientific program.

The administration of the team is critical. As discussed, often the team will be made up of experts, each of whom may normally be accustomed to playing a lead and independent role. In the parlance of current business organization thinking, the group is best viewed as a "virtual organization," operating as a single entity. The role of the leader, normally the banker, is to direct and orchestrate the efforts of the team. In addition, as the focus of the efforts of the team evolves from permitting to construction, to long-term maintenance, the composition of the team may change. It is the job of the leader to manage this change.

Having considered whether the banking enterprise generally makes sense and addressed the general nature of the capabilities and organization required, the banker is ready to begin to plan the enterprise in greater detail. As with most such planning, even the most basic assumptions of the enterprise will be questioned and refined from time to time as the effort goes forward.

In general, the banking enterprise normally can be divided into five phases (although they may in fact overlap):

- Feasibility assessment and preparation of a preliminary business plan
- Permitting
- Conservation work
- Sale of credits
- Long-term management of the conserved wetland/habitat

Phase I: Feasibility Assessment and Preparation of Preliminary Business Plan

The objective of the first phase, and arguably the most important element of the entire process, is the development of a preliminary feasibility assessment and business plan. The business plan will be revised further as the project proceeds, but it is important that it sets forth a realistic program for developing the bank. Topics addressed in the plan should include the following:

- Site selection
- Conservation program
- Schedule

- Financial plan
- Entitlements

Early Reconnaissance: Knowing the Territory

In undertaking the assessment and preparing the plan, as Harold Hill sang, "you gotta know the territory." There is no substitute for knowing the individual staffers in the various agencies, the land and resources involved, and the developers who will need the credits that are to be produced.

Because of the biological complexity and uncertainty of the undertaking, the banker must have a good sense of the ecosystem involved. How has the particular stream changed over the years? How has the population of gnatcatchers or vireos or butterflies varied? Can the coyotes be expected to ward off feral cats in the area? What is the history of the water table levels? How widely has the species been encountered?

In addition to understanding the biophysical environment, it is important to know the people involved. A good, trusting relationship with the staff of the agencies is important. They must have confidence in the bank sponsor, an attitude of support, and be willing to adapt to changing conditions (which are predictable). In a similar manner, knowing the various developers is important. Will they know to contact the banker when they need mitigation? What are the historic mitigation practices in the area? What are developers used to paying for mitigation?

Early reconnaissance is very important. This may occur when the bank sponsor has a specific bank site in mind or is seeking a bank site within a broader area. The sponsor should identify the "constituency" of the project that may include the agencies, owners of potential sites, real estate brokers, biologists, government officials, conservationists, and potential credit purchasers.

The banker's list of questions to be addressed to the constituency may be very broad. Among other things, it must focus on matching potential demand with the supply of mitigation. In general this requires a matching of biological attributes of lands to be developed and lands available for mitigation. In addition, it requires an ascertainment of the views of the regulatory agencies as to how they view and categorize the various resources, as well as the views of the conservation community.

To make matters more complex, the view of the agency staff members or conservation organizations may vary. As with taxonomy, there are "lumpers" and "splitters," that is, agency members that may view the resource as being made up of a large number of narrowly defined groups of plants and animals or, conversely, a small number of broadly defined groups. The factors underlying such a classification may range from geographic distance or watershed to specific species, subspecies, or distinct populations. Sources for this information may include agency staff, mitigation requirements in connection with past permits, and formal or informal policy statements of the particular office of the agency. This issue of classification is also of importance in the negotiation of the banking permits and agreements.

More specifically, the questions addressed to the constituency may include the following: What is the demand for mitigation? What has been the past development activity that has required mitigation? How have mitigation requirements been satisfied (thereby establishing a track record or trendline of experience)? Then, what development activity may be emerging or initiated that is likely to require mitigation? Agency records of permits and consultations can be reviewed and the staff can be asked for their input. Development permits issued to date in the area can be reviewed to help answer these questions. Inquiries can be made of the development community (or farming or timber industries if applicable), as well as real estate brokers, biological consulting firms, agency staff members, and the planning staff of the local agency. Local general plans and plans prepared by regional, state, and federal agencies should be reviewed, together with lists of sensitive and "endangered" or "threatened" listed species under federal and state laws. For example, the wildlife agencies in some states, such as California Department of Fish and Game, compile "natural heritage" databases that describe the wildlife resources and their range within the state.

What is the supply of land and opportunities available for mitigation? Would it be economic to conserve these lands or are these lands too valuable for other purposes or too costly to conserve? What is the nature and extent of the likely development service area (ecosystem, range, watershed) in relationship to the availability of mitigation sites? What is the price of land? How does the ultimate build-out or development within an area relate to the ultimate amount of mitigation lands available? Often biological consultants in the area have a good sense of mitigation opportunities in the area, as well as the price of lands. Real estate brokers and consultants often are engaged by developers to identify potential mitigation sites and have knowledge as to what mitigation opportunities are available.

Are there other mitigation banks operating in the area or in the process of formation? If so, have they been successful? What is their focus? In short, what will be the competition? What is the political geography of the area under consideration for the bank? Is there more than one local general planning jurisdiction involved? What is the make-up of the governing bodies of the agencies within the area? Often, for example, a local agency may not be supportive of mitigating, in another jurisdiction, the impacts of projects within its boundaries.

Site Selection

As in all real estate development, the most important determinants of success are location, location, and location. Factors to consider in selecting a site include the adequacy of legal title, biophysical characteristics of the site and surrounding areas, potential development (service area development requiring mitigation), potential competition, and the underlying institutional framework.

Title

A key initial step in analyzing the site is to review the state of legal title to the land. For this purpose a preliminary title report covering the lands should be obtained. If the land involved includes water-covered lands such as rivers, swamp lands, or tide or submerged lands, ownership may have been affected by past accretion, erosion, or evulsion or by adverse claims that are based on concepts generally not applicable to uplands, such as implied dedication prescription, public rights in lands adjacent to such waters, or the public trust. Normally, title insurers can be particularly helpful in assisting the banker in exploring these issues. Further, in selecting counsel for banks involving these types of lands, the attorney should have experience with respect to such special title issues.

As discussed at greater length in chapter 7, before expending significant amounts of money in exploring and establishing a bank, the banker should obtain assurances that it will be able to secure adequate title to the lands. Even in those instances where the banker owns the land, it should make sure that it will be able to assure the regulatory agencies, credit purchasers and others that the title is adequate. These assurances may be in the form of title insurance, a commitment by an insurer to issue title insurance, or a title opinion issued by a knowledgeable attorney.

Biophysical Nature of the Land

It is important to ascertain the physical nature and condition of the land for the purposes to be served, including the geology, the nature and condition of the soils, and the hydrology of the site, as well as related areas. In considering the development of wetlands, the nature of the soils and the current and anticipated depth of the water table is critically important. For example, in creating vernal pools, an impermeable soil stratum must be present. Likewise, in creating habitat for certain plants or animals, specific soils (e.g., clays for the San Diego thornmint) or specific host plants (such as lupine for the Mission Blue butterfly) must be present. While some of these basic elements may be recreated and maintained, the cost of doing so may render the bank unprofitable. Further, the development of an ecosystem that requires continued artificial support is generally frowned upon by the biologists and the regulatory agencies.

Service Area and Demand for Mitigation Credits

The success of a bank depends largely on locating it within a service area with adequate potential demand for mitigation. The determination of the service area is generally based on science and on the policies of the resource agencies. The service area may be defined by a watershed, a circle with a rather arbitrary radius, an area circumscribed by a local political boundary, or an ecosystem. It must, however, be adequate to provide sufficient development and related demand for mitigation. While the banker may initially determine what it believes the service area should be, this determination must be confirmed with the regulatory agencies in a preliminary manner at the time of the feasibility assessment and later in the bank MOA and permits.

To ascertain whether there will be adequate demand for the anticipated mitigation credits, the bank sponsor must conduct an appraisal of the amount and timing of anticipated development in the service area, the likelihood that the mitigation developed will match that required by other developers, the mitigation likely to be required by the regulatory agencies, and the amount of likely competition. These are very complex factors that can be subject to significant variation in particular circumstances. Current federal policies regarding mitigation may change. If the regulations regarding the amount of mitigation required for anticipated development were to change, the demand for mitigation would change as well.

Finally, as suggested earlier, often it is difficult to predict the amount of mitigation credits that may be available in the future. For example, as the southern California real estate market became increasingly depressed during the early 1990s, the amount of lands available for mitigation increased, all reflecting its market value for development purposes. Interestingly, this deluge of lands available for mitigation was encouraged by foreclosure by the various lenders, including the Resolution Trust Corporation, which as a matter of policy, were committed not to hold foreclosed-upon lands for long-term investment purposes. This further depressed land values in the region. In other cases, where the market for development is much thinner, it may be even more difficult to predict the value of the lands' mitigation banking purposes. For example, if a mitigation bank were to be established in a nonurbanizing area and were to catch on, other relatively low-value landholdings likely would be offered up for mitigation purposes as well. Thus, the mitigation banker must keep in mind the underlying "background" land values. It is for this reason that landowners are normally more willing to embark on mitigation banking for lands that they currently own since with a small investment or effort they may substantially increase the value of their holdings, as opposed to investing new capital to acquire lands for such purposes. It is wise, however, for the banker to consider linking the dedication of its lands to mitigation uses in case the bottom falls out of the market for mitigation credits.

This market risk can be further reduced as the result of broad-scale conservation plans (e.g., watershed, habitat, or special area management plans) that more specifically identify in advance the lands available for mitigation and those available for development. Of course, in this situation, as lands are identified under a plan or other arrangement as developable or nondevelopable, the scheme increasingly approximates a transferable development rights (TDR) program.

Scale

The lands selected for acquisition must be large enough to create sufficient credits to pay for the acquisition of the land, as well as to finance banking operations. Further, the size must take into consideration the questions asked earlier regarding the amount of mitigation activity projected, as well as how many other banks are or will be in operation. Usually, lands with the greatest restoration potential are the best candidates for acquisition. However, other options are available. Establishing a bank by creating wetlands where none existed before is the riskiest option, given the high rate of failure of wetlands creation. Restoration coupled with enhancement and protection of existing systems affords the best mix of banking opportunities, provided that the site fits with the broader ecosystem.

Resource Agency Preferences

Because biological resources often appear to be very complex and subject to varying scientific valuations and views, it is extremely important, as suggested earlier, that the banker's analysis of the existing and prospective resource values involved in the particular bank are shared and confirmed by the regulatory agencies with jurisdiction over the bank. Ideally, this confirmation can be provided by a broad-scale conservation plan. In the absence of such a plan, the initial assessment of the site should be confirmed to the extent possible by conversations and correspondence with the staffs of the local, state, and federal agencies with jurisdiction and in the Memorandum of Agreement (MOA) or permits establishing the bank.

If the bank is not located in an area earmarked in a regional or local general plan or conservation plan, the agencies likely will require that mitigation occur at locations that more closely satisfy the strategic objectives of their plans and determinations. For this reason, it is essential in selecting possible banking sites that the agencies are consulted early in the planning process to ascertain the specifics of their formal or informal conservation plans or strategies. The most successful banks focus on these needs and form, in essence, a partnership with the resource agencies. The benefit to the agencies is the ability to achieve the goals of their plans without spending their limited resources and without being required to go through complicated governmental acquisition processes. This is one of the potential benefits of embedding mitigation banks in broader planning frameworks, such as watershed, habitat conservation, or special area management plans.

Surrounding Area

Other considerations in selecting a site include the past, current, and prospective uses and character of the surrounding area, including, the vegetation, soils, and topography; the percent of the watershed lying outside the bank; and the proximity of the site to the urban environment. In this regard, the banker can ascertain the need for security, the maintenance of various proposed improvements and facilities, possible emergency responses, exotic species control, protected species management, and, where appropriate, fire management and/or control, public recreation, timber, grazing, and/or game management. Proximity to the urban environment will likely result in increased vandalism, greater problems with invasive exotic and feral species, more difficulty and less flexibility in use of prescribed burning, increased air and water pollution impacts, and greater demands for public access. A major cause of bank failures is the proximity to incompatible land uses. For example, a 100-acre forested wetland surrounded by a parking lot will succumb eventually to the effects of the destruction of the upland watershed.

Obviously, if the area encompassed by the mitigation bank does not contain the surface rights necessary to assure the hydrology of the wetlands, considerable manipulation may be necessary to maintain wetland functions over time. This translates into increased maintenance costs. In addition, as the science and art of wetland enhancement, restoration, and creation are relatively new disciplines, the economics of the bank must reflect the attendant uncertainty, if these elements are included in the conservation plan for the bank.

Community Care and Commitment

The siting of a bank should also take into consideration whether it will be met with support from the local community. Banks established as a valued part of a community's desired and necessary conservation infrastructure (often as reflected in the community's general plan) are much more likely to be monitored and cared for over the long term (see, for example, the West Eugene case study). Likewise, a community-related public or nonprofit conservation entity that undertakes to manage a site will probably enjoy greater long-term support than will most developers or private mitigation bank operators. For this reason, a banker may well consider the bifurcation of the enterprise, with the long-term management and operation of the conserved habitat being undertaken by a public sector agency or nonprofit organization. Further, and beyond the scope of this book, the conserved habitat, particularly where it is operated by a community-based nonprofit organization, may become the focus of far broader conservation activities, including, for example, educational programs and recreational uses.

Local, State, and Federal Governance Framework

A mitigation bank will be more successful where (1) the background value (that is, value for other uses) of the lands is at least equal to the value for mitigation banking purposes where the lands are to be acquired and is less than this value where the lands are already owned; (2) the cost of improvement for the banking use before the sale of credits is relatively small; (3) the individual projects within the service area each, individually, would result in a relatively minor impact to the ecosystem, but collectively have a significant effect (increasing the acceptability of the loss of some habitat in return for off-site mitigation); (4) there is significant, ongoing development and demand for mitigation; and (5) there is clear multiple agency collaboration and agreement upon the desirability for, and factors to be addressed by, a mitigation bank.

Preparation of Conservation Program

The second most critical component of the feasibility assessment is the development of a preliminary conservation program. This program must go hand-in-hand with the financial assessment and planning for the effort. The elements of such a conservation program include the following:

- A baseline survey of biological resources
- Proposed enhancement, restoration or creation objectives and work—the conservation work
- Monitoring program
- Preliminary budget and anticipated cash flow
- Provisions for adaptive management in the event of unforeseen circumstances

Determining the Credits

The basis for the determination of the value of the mitigation credits should be outlined in conjunction with the conservation program. This valuation is perhaps the most critical element of the formula that determines the economic success of the mitigation bank. The valuation is often articulated as a conversion rate between the impacts or conservation losses to be mitigated by bank credits and the conservation benefits to be provided by the bank: 1:1, 1:2, or more. It may be articulated as a standard—wetlands that are as biologically productive as those impacted by development—and may be qualified by geographic boundaries (e.g., the same watershed) and distances (e.g., 15 or 40 miles), as well as requirements related to the kinds of species and habitat types and characteristics that must be present. In some cases, credit may be provided for adjacent lands of a different habitat type (e.g., as buffer or supporting lands). Credit for these adjacent lands may be important because in acquiring habitat, it is not always possible to acquire only a portion of a parcel.

A few programs have been developed to compare the relative biological values of different lands. These include the Wetland Evaluation Technique (WET) and Habitat Evaluation Procedure (HEP) programs. It is very difficult, however, to capture in a fixed program all of the factors underlying the determination of value. For example, valuation protocols that consider only biological factors may not properly include consideration of the risk of loss of similar habitat because of a lack of title or regulatory controls (an issue that is addressed by a "gap" analysis that identifies the resource and then determines the portions of the resource that are relatively protected legally and those that are at risk). Further, the encouragement of mitigation banking may have such a positive effect on conservation that it may be worth the risk of defining the acceptable habitats more broadly or by increasing the geographic extent of the service area. Without question, the regulatory agencies tend to narrow the breadth of possible exchanges where the bank is established within a milieu of project-by-project permit determinations. On the other hand, where the bank is established as part of a broader conservation planning effort, the agencies are inclined to broaden the possible exchanges. When proceeding on a project-by-project basis, the agencies have less assurances of the cumulative result of many individual transactions, while under a broad conservation plan, they know the ultimate conservation levels and have control of the process. The preliminary conservation plan should articulate the logic of both the conservation strategy within the conserved habitat of the bank, as well as the relationship of the conserved habitat to the broader ecosystem.

Legal Elements

Finally, the conservation program should outline the legal elements necessary to implement the program, as discussed in greater detail in chapter 7. Normally, this would include provision for an implementation agreement, ideally entered into with all of the agencies having jurisdiction, and setting forth the undertakings of the parties to complete the conservation work, to maintain the habitat, and to honor the credits provided. Topics should include assurances regarding the proper completion of the work (which can be addressed through performance bonds) and the character of the entity that will hold the conserved habitat over the long term, as well as the funding of that maintenance.

Schedule

A major source of cost overruns and mitigation banking failures is the inability to accurately assess the time necessary to complete the bank and sell the mitigation credits. Accordingly, the banker must accurately and conservatively estimate the time necessary to complete each step of the process. This should include, among other things, addressing title issues, conducting necessary biological surveys and studies (sometimes requiring surveys during a specific and limited season or over several seasons), negotiating with the various agencies, completing the conservation work, and marketing the credits. A time schedule is critical and should be prepared and monitored as the effort unfolds.

Financial Plan: Budget, Cash Flow, and Projected Profits and Losses

For the banker, the bottom line is expressed in dollars. Thus, a financial plan must be prepared as a part of the assessment. It should include a budget and projected cash flow and proforma profit/loss statement for the anticipated life of the banking enterprise.

The financial plan must reflect the estimated revenues and expenses of the enterprise, adjusted based on the time anticipated for each step. Because of the complexity and uncertainty of the effort, it is difficult to accurately estimate the amounts involved.

Unlike a normal business plan, the economic feasibility analysis must reflect the role and policies of the public sector agencies, such as the amount of time that will be required in permitting, the definition of the service area, and the mitigation ratio. In addition, the public sector can choose to subsidize, in effect, the mitigation banking enterprise. For example, a public agency, such as a city or conservancy, may be willing to undertake the long-term maintenance of the conserved habitat or may be willing to contribute to the mitigation burden of the developer by lowering the number of credits that must be acquired. Any such effects must be taken into consideration in the economic analysis. In considering such policies, the public agencies should realize that in undertaking the long-term management of the conserved habitat, the banker's analysis will not only reflect a reduction in the actual cost of the management but will also include a reduction in related budgetary allowances. The revision in the budget projection could well be greater than the actual costs involved.

As suggested earlier, the complexity of the undertaking, as well as the non-free-market aspects of wildlife conservation and governmental processes, make the process particularly susceptible to delay. The banker should consider ways to build into its financial projections adequate reserves for delay and to otherwise reduce its risk. A portion of the risk of delay could be shifted to the owner of the land (e.g., by negotiating a price based on the achievement of certain timing objectives or by phasing a series of purchase options to correspond with increasing demand).

Permits and Entitlements

The preliminary assessment should include a careful listing of the various permits and entitlements required, including, for example, Section 404 permits under the federal Clean Water Act (and Section 401 certifications) and Section 10(a) permits under the federal Endangered Species Act, as well as state and local permits and approvals. In addition, it should articulate the "logic" to be used in addressing the various permitting requirements.

The assessment should include a preliminary listing of the surveys and other engineering work that may be required by the agencies, the service area, the conservation work, and, as discussed, the nature of the credits to be provided (including the basis for determining the economic relationship between the mitigation credits to be provided by the bank to the development impacts mitigated).

Establishing the Banking Entity

At some point during Phase I, the banker will begin to establish the institutional framework for the bank. Depending on the nature of the bank (i.e., for-profit, nonprofit, public sector, or single-user), this may occur in one or more steps. For example, all elements, from permitting to long-term maintenance, comprising a bank established by a utility may be organized as a division of the utility corporation. Alternatively, an entrepreneur may form a corporation to obtain the permits and sell the credits, while contemplating that it will ultimately convey the conserved habitat to a public agency or nonprofit corporation for long-term maintenance, with the provision that the banker will provide active day-to-day management as an independent contractor.

A detailed discussion of the appropriate form of business organization for a bank is beyond the scope of this book other than to make the general observations made in the early part of this chapter and in chapter 7. In connection with entrepreneurial banks, the legal structure of the bank probably will be a for-profit or nonprofit corporation. Where the bank is a nonprofit organization, it may be a land trust or conservancy. In some cases, the bank may be operated as part of a public agency, such as the Riverside County Habitat Conservation Agency discussed in chapter 7 and later, and may either sell mitigation credits with respect to a specific parcel of land or as discussed in the Introduction/Overview may simply charge an impact fee based on the estimated cost of providing the necessary mitigation.

Any of these organizations can be utilized to obtain title to the lands, obtain the necessary permits and approvals, perform the mitigation work, sell the mitigation credits, and manage the conserved habitat for the long term. However, it may be desirable to bifurcate these functions, with the long-term ownership and/or management of the conserved habitat being undertaken by a separate entity. This is not unlike seniors' housing projects and cemeteries in which the separation reflects the sharply different functions of development and initial marketing and ongoing, long-term management. The management entity may be a conservancy or land trust established on a project-specific basis, a public agency such as the state fish and wildlife agency, or a local agency. As discussed in chapter 7, some have suggested that a special class of corporations, similar to cemetery organizations, be established for the purpose of holding such conserved habitat, with appropriate limitations on their powers established by state law.

The main reason for such a bifurcation is that the banker may be an entrepreneur that is primarily interested in profiting from the sale of credits and has no interest in attempting to profit from the long-term management of such lands. While historically the management of such lands has not been viewed generally as a profit-making activity, that view may be changing, particularly where the bank has been endowed in advance.

A variation to this approach is for the long-term management of the conserved habitat to be further bifurcated with the lands being owned by one entity, normally a nonprofit corporation (land trust or conservancy) or public agency, and operated day to day by a for-profit contractor. The choice of the appropriate structure will depend on a variety of factors. For example, while it may be appropriate to engage a contractor to operate the conserved habitat under some circumstances, if there is a potentially interested local community in the vicinity, a professional contractor may discourage volunteer participation and interest. In such a case, it may be appropriate to consider the formation and use of a locally based land trust or conservancy to manage and operate the conserved wetlands/habitat.

The Conservation Program

The focus of the conservation plan is the conservation program—the creation, restoration, enhancement, and preservation of the wetlands/habitat. Based upon the initial biophysical surveys and studies, the conservation program is much like any other plan of development. It may be comprised of separate, but integrated, plans for soils and topography (grading), hydrology, vegetation, infrastructure, and services and structures, as well as for the long-term maintenance and management of the resulting conserved habitat. These plans must be carefully and thoroughly coordinated within the consultant team.

The program should also provide for monitoring and unanticipated events. These elements of the program should be carefully coordinated with the phasing of the availability of credits, as well as the other elements of the conservation plan. As with other elements of the enterprise, the regulatory agencies should be consulted. In particular, the local agencies with general planning jurisdiction will have a particular interest in making sure that the program complies with local planning and building regulations, as well as public health and safety requirements. For example, where wetlands are to be created and maintained in proximity to urbanized areas, an element of the program should address mosquito abatement and control.

Acquiring the Bank Lands

Before committing significant funds to a specific bank site, the banker must obtain sufficient title commitments to assure that it will be able to complete the bank; that is, to complete the conservation work and provide title to the long-term management entity. While this may be accomplished by the acquisition of fee title or an adequate conservation easement at an early point, alternatively, the banker may acquire an option to purchase the lands (which may provide for acquisition in phases), or a lease with an option to purchase. This allows the banker to hedge its bets, to proceed to acquire title only if the permitting effort is successful. It also allows the banker to ask the landowner to share the risk of delay by providing for reduced interest during such extended periods.

Title insurance or an opinion of title is important to the banker, as well as to the agencies. It should be obtained at the time of the lease, option, or acquisition of title. Of particular concern are exceptions to title that may allow others to use the conserved habitat (e.g., for mining, oil development, or roads) or that would affect the boundaries. For water-covered lands, the location of boundaries can be difficult to ascertain and often title is encumbered by public trust and other interests. The banker should take care to engage legal counsel that is familiar with this issues.

Phase II: Obtaining the Permits and Other Entitlements

Draft MOU

As the feasibility assessment phase evolves into the entitlement phase, the banker should begin the preparation of a draft Memorandum of Understanding (MOU) or discussion outline for use with the agencies. This draft MOU or outline should address all of the points upon which the banker ultimately desires to receive formal or informal assurances from the agencies as part of a Memorandum of Agreement (MOA), including an outline of the entitlements as discussed earlier, anticipated surveys and studies to be required, the legal elements of the conservation plan (including an outline of the MOA), and the schedule. Increasingly, regulatory agencies are working together, in a collaborative mode, with applicants. The banker should encourage such an approach. In some cases, where the mitigation bank is unusually large or part of a broader conservation planning effort, the banker should consider using a neutral facilitator to design, orchestrate, and direct the entitlement process.

Final MOU

As agreement with the agencies is reached, the process MOU should be revised and finally signed or initialed by all parties. While the MOU may not be legally enforceable and may expressly provide that it is not a legally enforceable document, it should be viewed as providing the moral commitment of the signatories to process the entitlements in the manner set forth in the MOU. A major reason for not being concerned about the enforceability of the MOU is the time that would be required to accomplish the necessary legal review and approval if it were to be enforceable. In some cases, such as a very large scale bank, or a bank embedded in a broader conservation planning effort, it may be appropriate to develop and expressly provide for a legally enforceable process MOA.

While the MOU may not be enforceable, it can provide remedies for a failure to perform. For example, it could provide that if the time schedule is not met or a certain issue is not resolved to the satisfaction of all of the parties within a specified amount of time, the matter will be elevated to a higher level in the agency for review. This is consistent with trends in the agencies toward greater vertical management of issues. Increasingly, higher-ups are willing to become involved in issues of greater moment or of a precedent-setting nature. In some cases, fixed review points with these higher-ups can be established as part of the time schedule. Arrangements for such vertical management normally require that the process is designed with the participation from such higher-ups from the outset. For a number of reasons, some of which are quite reasonable, the lower tiers of an agency are rarely prepared to provide for such an appeal process to their superiors.

Negotiating the Process MOU

As suggested, several points in negotiating the MOU should be emphasized:

- The use of a draft MOU, even in a very crude, early form, is helpful.
- Collaboration by the agencies is appropriate and should be encouraged.
- The appropriate level within the agencies of the negotiations will vary depending upon the size of the project, its precedent-setting nature, and its complexity.

At the outset, a very informal understanding should be reached regarding the process (manner, timing, issues, etc.) for the completion and signature of the formal process MOU. The agencies involved have not had a tradition of working in this manner, although this is changing. Accordingly, resistance may be encountered to the idea of developing a process MOU. Some federal agencies simply are not familiar with the process MOU. Within state and local agencies, there is a greater likelihood that they are simply not used to working in this manner. Various techniques can be employed to overcome this resistance. For example, the federal agency with this expertise can be encouraged to take a greater lead role.

The conceptual basis of the collaborative planning approach is the more horizontal, collaborative management structures first pioneered in the private sector and characterized by "partnering" and "management by value and principle." It assumes a more visible, on-the-table, multiple constituency involvement in these processes, with greater use of creative agreements and accountability (involving more basic elements of respect, consideration, and fairness) for all of those concerned. The collaborate development of the process MOU is the first step in the program that will ultimately result in the bank and all that is envisioned as a part of the bank. The entire process should be viewed as a collaboration by a team of interests and agencies seeking to orchestrate their work together to maximize each of their individual interests. The MOU is the "score" of that effort, noting the timing and performance required of each participant. In some cases, the score will be more open, requiring that issues be addressed and resolved. In other cases, the MOU will be very specific; for example, in connection with the legal elements of the arrangement for the long-term management of the conserved habitat.

Phase III: Implementing the Conservation Plan

With the necessary permits and entitlements in hand, the banker is prepared to implement the conservation plan. The plan provides the basis of the permits and may be accompanied by an implementation agreement. As discussed in chapter 7, the agreement sets forth the legal framework supporting the plan, providing for funding, security, amendments, and remedies.

The lead in the orchestra/team moves to the biologists and to the grading and building contractors and landscapers. They must make the plan come alive. Creating or restoring a wetland is becoming easier, but it requires careful attention to details, adaptation to unforeseen circumstances, and patience.

The agencies should continue to be viewed as a part of the collaboration. Monitoring reports will be a part of the approved conservation plans, and changes in the plan are predictable. Accordingly, it is important to continue to nurture and tend the relationships that were created during the permitting phase.

What Can Go Wrong?

There may be a Murphy's Law just for mitigation banks. Some common problems include misapplication of fertilizers, using seed sources contaminated with exotic weed propagules, and predation by herbivores. Geese have been known to denude an entire newly planted wetland site in less than a day. More commonly, human errors in the design or implementation are to blame for the poor performance of some mitigation banks. At one site, crews planted an intertidal wetland in the evening when the tide was out and installed the plants upside down. In another, clay layers in the soil were not identified, creating unanticipated drainage problems and higher construction costs. All this and more has gone wrong in the past. Even the most conscientious, well-laid plans can fail.

In addition to unplanned occurrences, there has been a historic tendency for the entrepreneur, after it has obtained its profit objective, to lose interest in permit compliance. One only needs to drive through the strip-mined counties of southern Ohio to see the results of inadequately assessing the risks of noncompliance. Decades after the coal was removed, thousands of acres of land are still unreclaimed. Hundreds of miles of streams continue to be polluted by acid mine drainage. This occurred because, at one point in the history of the industry, reclamation was not required or it was cheaper to default on restoration, forfeit the minimal bond, and start mining again under a new company name.

Abandoned commitments have been a common problem in the history of wetland mitigation as well. In many cases, the failure of wetland mitigation projects stems from a failure of institutions as much as a failure of mitigation know-how. Unless the regulatory community is prepared to adequately articulate and assign risks, mitigation banking will suffer the same fate. Bonding is a common, if imperfect, solution to this problem.

Many wetland creation projects failed because permittees tried to cut corners to save money. Well-planned mitigation projects are expensive. Mitigation banking, too, will fail if it is viewed as a cheap alternative to on-site mitigation. The following list provides a partial listing of what has gone wrong with a number of wetland mitigation projects in the United States over the past decade.[1]

- Grading plans not followed
- Tidegates incorrectly installed; planted cordgrass died

- Substrate not excavated deep enough
- Dump site discovered during excavation
- Poor or no monitoring data
- No budget to make midcourse corrections or fix postconstruction problems
- Insufficient operation and maintenance of site
- Soils not analyzed
- Soils inappropriate for restoration because of type, compaction, salt layer, woody debris, and so on
- Siltation and erosion not considered
- Outfall channel filled with silt
- Culverts were too small or blocked with silt, vegetation, or debris resulting in stagnant pools, alga blooms, lack of tidal flushing
- Personnel changes; no transference of data or procedures
- Site was overengineered
- Siltation of tidal entrance
- Planting done at wrong elevation
- Failure to analyze groundwater or freshwater inflows
- Improper site conditions for restoration
- Uncontrolled invasion by exotic plants and animals
- Overzealous weed control
- Destruction of vegetation or substrate by catastrophic events
- Failure to protect site from off-site or adjacent impacts
- Failure to maintain water levels
- Insufficient irrigation for plant establishment
- Contamination from unanticipated urban storm run-off
- Human intrusion and vandalism
- Revegetation done at wrong time, with wrong plant, or with wrong species leading to unnatural appearance or failure
- Frequency of inundation; too much or too little
- Vague or undefined project goals and objectives
- Undefined project success criteria

Upland mitigation is simply too new to have a history of failures. Further, it tends to be comprised more of preservation of the existing habitat. An exception is the San Bruno Mountain conserved habitat, established under a Habitat Conservation Plan (HCP) and federal endangered species incidental take permit. Under this HCP, the active eradication of gorse, an invasive exotic plant species, has been a major focus. The objective is the restoration and maintenance of the grasslands that are the habitat of the mission blue butterfly (a federally listed endangered species), as well as other species. Undertakings in the accompanying implementation agreement, including monitoring, have assured that this effort has progressed over the past decade in which the HCP has been in effect.

Phase IV: Marketing and Conveying the Mitigation Credits

The bottom line for the entrepreneurial mitigation banker is profit from the sale of the credits. Normally, under the MOA, the conveyance of credits will occur after the regulatory agencies have certified in writing that the mitigation bank has achieved the agreed-upon objectives for mitigation. (Normally, mitigation banks must demonstrate their success at restoring and creating wetlands or habits in advance of permission to sell credits, unlike on-site mitigation procedures.) Certification can be accomplished by a written confirmatory letter or a notation in the records/accounts of the bank. For purposes of protecting against the double sale of credits, the MOA should spell out the records/accounts that can be relied upon by a potential credit purchaser in determining whether credits are available and have not been previously sold and that no sale is effective until entered in the records of the bank.

The banker, however, can enter into contracts and options to sell and purchase credits, obtaining up-front cash payments that can be used to fund the bank. The arrangements can vary significantly, ranging from simple payment on the provision of the credit, to prepayments, to commitments, for example, by the credit purchaser to provide further funding for maintenance. In structuring these more complex arrangements, the laws of the particular state with respect to the regulation of real estate sales and securities should be carefully reviewed.

As discussed in chapter 7, the MOA can provide for the phasing of mitigation credits, based upon the achievement of various milestones in acquiring, restoring, or creating the resource. Completion of the subsequent phases of the bank can be assured by completion or performance bonds.

For more complex conservation programs, there may be some question as to whether the milestone has been reached. As a result, the regulatory agency may balk at providing the written certification necessary to proceed with the next level of credit sales. A carefully drafted MOA will help the parties get past potential impasses. It should provide clear standards and the use of mediation or even arbitration to resolve such issues.

In consummating the credit purchase, the purchaser may desire a conveyance of the mitigation credit that is signed by the banker as well as the regulatory agencies, confirming that with the payment of the agreed-upon amount, the mitigation credit will be applied against its mitigation obligation under its specific, for example, Section 404 or Section 10(a), permit.

Marketing the mitigation credits may take a variety of forms depending on the circumstances. In some cases, the banker is known, or makes itself known, to the development community. In other cases, an agent may be employed. The agent may be a real estate agent that normally brokers land for development purposes.

Phase V: Long-Term Maintenance of Conserved Habitat

The significance of perpetual stewardship of banks cannot be overstated. The local, state, and federal communities are often relying on this undertaking to conserve the biodiversity of the Nation. As discussed briefly earlier, the banker may choose to focus on the creation of the bank, the completion of the conservation work and the sale of credits, and delegating the long-term management of the conserved habitat to a public agency or nonprofit entity. The delegation can be accompanied by an endowment that will provide sufficient current income when conservatively invested (e.g., sufficient to provide a 6 percent per annum return) to manage the conserved habitat. Alternatively, the management of the bank may be assumed as a public responsibility and funded by taxes, assessments, or other public sources. Any such public contribution to the operation of the bank should be taken into consideration in determining the value of the mitigation credits. Such arrangements can be anticipated more often in connection with broad-scale conservation planning efforts.

Following the initial conservation work, banks can be expected to minimize their management requirements once the conservation objective has been achieved. True success can be defined as that point "when the site reaches a self-maintaining state . . . with a high degree of persistence and resilience to natural and anthropogenic disturbances and does not require external management inputs to stay viable."[2] In the highly fragmented, human-influenced landscape of North America, it is difficult to envision many places where no external management inputs are required to maintain the viability of a mitigation site. Even in largely intact designated wilderness areas in North America, some management is vital to maintain the resources.

In most areas, active management will be necessary to maintain the biological integrity of the site; some examples include fire management in fire-adapted systems or control of exotic plants and animals. All mitigation sites will require some protection from the threat of vandalism, poaching of plants and animals, dumping, and boundary incursions. Additionally, all sites will need to have a long-term "institutional memory." As discussed in chapter 7, legal mechanisms must be in place to guarantee that as lands change hands, subsequent owners and managers will not convert the site to other uses. As suggested, conservation easements can be an effective tool to protect the character of the conserved habitat; however, they are only as effective as the efforts made to monitor and ensure compliance.

Depending on the complexity of the hydrologic issues, some sites will have dikes, weirs, pumps, or other infrastructure that will require maintenance. In bank sites where these types of engineered structures are a part of the management regime, one also has to plan not only for routine maintenance but for design flaws and other human error. It is only reasonable to expect some error and to build in resources to address unanticipated problems.

The management of natural resource lands is a process, not a product. There may be products derived (mitigation credits, waterfowl, improved water quality, etc.), but the land itself is organic. As such, a manager will need to respond to the evolving and ever-changing nature of the system. All sites will be subject to natural events (floods, fires, drought, insect infestation, etc.), the effects of which could be severe but largely unpredictable. The agreements developed for these banks need to recognize the inevitability of natural, unpredictable events. They need to also allow for some flexibility in deliverables. At one mitigation site, the activities of a beaver altered the hydrology of the site and produced a different type of wetland system than was called for in the permit. Given the unpredictability of such natural events, the idea of success and failure for a given area will need to be responsive to the patterns of natural change. Some systems must change to persist. As D. E. Willard and J. E. Klarquist point out, often we attempt to recreate or preserve a specific wetland type with a particular species mix and precise geography. Now we accept that wetlands are living systems and some types do change. They grow, change species, and become other systems. Yet we often prescribe conservation plans that dictate constancy and attempt to construct a particular kind of wetland or habitat in place forever.[3]

Funding Long-Term Management

One of the most critical issues in the establishment of the bank is the provision of an assured source of funding for the long-term management of conserved habitat. This is similar to the requirements of operating a cemetery and is common to more generalized mitigation requirements that contemplate the long-term management of the conserved habitat or resource. The elements of that funding include determination of the cost, identification of the source of funding, and, assurances that the funds will be provided and expended as contemplated.

Determining the Cost

Among the planned or established banks there exists considerable variation in how long-term funding needs were determined. In several banks in California and in one private bank in Georgia, there is no funding source identified for management of the site after the restoration goals have been met. The assumption is that the bank operator (public agencies in the California cases and private-fee simple landowner in Georgia) will be able to absorb any long-term management needs within their existing activities (e.g., flood control, parkland, and open-space maintenance). As most federal and state agencies face continued cutbacks and downsizing, this approach is less likely to produce the desired conservation objective. This strategy has resulted in at least one known failure in a DOT mitigation bank in Louisiana.

In most mitigation banks, estimates of the cost of long-term management were not based on the actual goals of the project and the true costs of implementing the activities necessary to achieve those goals. Instead, the size of the fund often was determined arbitrarily. At the Florida Wetlandsbank (see case study 2), the banker was required to create an endowment of $1,000 per acre to manage the bank in perpetuity. This figure apparently was derived from management costs established for a site over 20 times larger, containing very different habitats and located over 200 miles away. Another state agency that has developed mitigation banks to offset impacts to protected upland species, has established the figure for long-term management endowments for these banks as 15 percent of the cost of acquisition.

For most long-term management arrangements, no assurance is given that they will provide sufficient resources to manage the areas over the long term. This lack of attention to providing for the true costs of long-term management will result in the failure of most mitigation banking projects.

The state of Florida has addressed this concern in the Department of

Environmental Protection's mitigation rule approved February 1994. Rule 17-342.100 requires that mitigation permit applications include "The cost estimate for the long-term management of the Mitigation Bank [that] shall be based on the costs of maintaining and operating any structures, controlling nuisance or exotic species, fire management, consultant fees, monitoring activities and reports, and any other costs associated with long-term management."

Unfortunately, reliable sources for some of these costs are not readily available. Most similar land management activity has been conducted by government agencies, and their costs are not easily extrapolated to the private sector. For example, the State of Florida's Department of Agriculture provides some prescribed burning services to private landowners in Florida. The service is primarily geared toward fuel reduction burning, is limited to the availability of staff, and is delivered on a first-come, first-served basis. The state charges $12.50 per acre for this service, which is below the actual costs. Since this service is subsidized by the state, a budget based on this figure would underestimate the true costs. A 20-acre burn in an urbanizing area with major smoke management issues could never be accomplished for $250 in the private sector.

The following list provides some information on sources of cost data for construction-related costs that would be germane to many restoration tasks needed on bank sites.[4] Unfortunately, the costs of some of the potentially required activities are not well documented in these sources for all regions of the country. For these activities, the cost might best be determined by estimating the types and numbers of people required to accomplish the task, multiplying these individuals by the time spent and the average cost per hour, and adding to this labor figure an analysis of equipment and material needs.

- *Means Site Work Cost Data* (R.S. Means Co., Inc.; published annually).
- *Dodge Guide to Public Works and Heavy Construction Costs* (McGraw-Hill; published annually).
- *National Construction Estimator* (Craftsman Book Co.; published annually).
- *Rental Rate Blue Book for Construction Equipment* (Dataquest, Inc.; published annually).
- *Construction Equipment Ownership and Operating Expenses Schedule* (U.S. Department of Army, Corps of Engineers; updated occassionally).
- *Labor Contacts Manual* (various trade associations; monthly/quarterly).
- *Engineering News Records* (McGraw-Hill; published weekly).

Figure 9.1 provides a list of the types of things that need to be considered when determining the long-term management budget for a mitigation bank site. Obviously all of these activities will not be necessary on all sites, however, every bank should be accompanied by a similar project cost analysis.

Securing the Funds

Once long-term management costs have been estimated as closely as possible, a mechanism for securing the funds must be established. Four basic approaches are used.

The first, and perhaps simplest, is the establishment of a perpetual endowment. In this scenario, the estimated endowment needed is apportioned and allocated to the credits available for sale by the bank. This amount is then deposited in a trust account as the credits are sold (or in advance of sale, as a cost of the bank), the interest from which will be used to manage the bank by the long-term designated manager. Because banked mitigation wetlands are designed to compensate for a wetland that conceivably would have existed for decades—if not in perpetuity—it is not unreasonable from a conservation perspective to require a perpetual-care fund. Further, the calculation of the amount of the endowment must take into consideration the effect of inflation.

Another method that has been used is to establish a designated period of management liability (e.g., generally defined as the time needed to demonstrate stability and ecological success) and create a "wasting" endowment. These types of endowments are calculated to meet the needs of the bank throughout the term of the agreement and, through a combination of capital and interest expenditures, be depleted at the end of the term. These types of arrangements are most likely to be successful if, after the success criteria have been met, the probable long-term management needs will be minimal.

The third approach is to enter into a contract between the banker and the entity requiring mitigation credits for an annual management fee for some designated time or until some success criteria have been met. This allows for less up-front investment by the credit purchaser but provides less security for the banker. This type of arrangement is used most effectively if the permit liability rests with the purchaser or if there exists a bond or other insurance should the purchaser fail to provide the required funding.

The fourth strategy is to convey the bank to a public or nonprofit conservation land-managing entity and rely on that organization to secure public funds (e.g., through taxes and assessments) or private contribu-

ACTIVITY	FREQUENCY	PERSONNEL NEEDS			FIXED CAPITAL	
		STAFF TIME	AVG HR WAGE	TOTAL COSTS	ITEMS	PURCHASE/DEPRECIATION RENTAL
1. SECURITY						
PATROL: ARMED						
PATROL: UNARMED						
FENCING: MAINTENANCE						
SIGN: MAINTENANCE						
SIGN: REPLACEMENT						
2. GENERAL MAINTENANCE						
ROAD GRADING/REPAIR						
TRAILS						
BUILDINGS						
MOWING						
DITCHES						
CULVERTS: CLEANING						
CULVERTS: REPLACEMENT						
DIKES MAINTENANCE						
CONTROL STRUCTURE — REPAIR AND MAINTENANCE						
EQUIPMENT: REPAIR						
3. EXOTIC SPECIES CONTROL (ASSUMES MAJOR INFESTATIONS HAVE BEEN CLEARED)						
MONITORING						
REMOVAL: PLANTS						
REMOVAL: ANIMALS						
4. FIRE MANAGEMENT (ASSUMES SITE FIRE PLAN IS COMPLETE AND PRIMARY LINES ARE IN PLACE)						
UNIT PLANNING						
PRIMARY CONTROL LINE MAINT.						
UNIT PREPARATION: MOWING, DISKING, AND RAKING						
UNIT IMPLEMENTATION						
FIRE SUPPRESSION						
NEIGHBORHOOD OUTREACH						
FIRE EQUIPMENT: MAINTENANCE						
FIRE EQUIPMENT: REPLACEMENT						
5. MONITORING (ASSUMES ALL PLANS ARE COMPLETE AND TRANSECTS AND WELLS ARE IN PLACE)						
VEGETATION TRANSECTS						
WATER LEVELS						
WATER QUALITY						
PHOTOSTATIONS						
WEATHER STATION READINGS						
DATA MNGMT INTERPRETATION						
MAINTAIN EQUIPMENT						
MAINTAIN TRANSECTS						
6. GENERAL ADMINISTRATION						
FINANCIAL REPORTS						
PROGRESS REPORTS						
PERSONNEL MANAGEMENT						
PERSONNEL TRAINING						

Figure 9.1 Recurring Management Costs Worksheet

tions to cover the management of the site. In California, in connection with habitat conservation planning under the federal and state endangered species acts, private assessments on adjacent development as well as benefit assessments under a state-enabled Habitat Maintenance Assessment District are available to provide such funding. Clearly, this source is less assured than the up-front endowment. Although government entities are more likely to exist in perpetuity, funding problems and financial shortfalls for resource management are also a perpetual problem for these entities. And private nonprofits are subject to cycles of boom and bust that affect philanthropic giving and interest.

Banking arrangements relying on public agency or nonprofit organization management over the long term should be carefully reviewed by the permitting agencies. If restoration and management are extremely complicated and/or experimental, it is likely to be unrealistic to assume that the long-term manager will be able to secure the needed funds to continue the work or repair any unforeseen problems. Examples of this arrangement have led to uncompleted projects languishing while the parties were locked in conflict over culpability. Similarly, if public-sector or nonprofit organization-managed banks are not based on an accurate estimate of the real costs of the bank, they not only risk undermining the ecological goals of conservation regulations, they also subsidize the costs of development and subvert the private market for mitigation banking by introducing below-cost competition.

Monitoring Success

All mitigation banks should adopt some performance criteria defining the conditions that must be met before the mitigation will be deemed successful and credits can be conveyed. Some of these will be as simple as ensuring that the preserved wetland has not been converted to other uses and the vegetation is not dominated by exotic species. Others will be highly complex depending on the criteria established during the permitting process. Whatever the criteria, monitoring is required, including reports to the regulatory agencies and confirmation of goals achieved.

Establishing agreed-upon standards and procedures is a critical component of the conservation plan. To an extent, this monitoring will assist the banker to know whether engineered devices are functioning and stable, the hydrology is consistent with what was predicted or expected, the vegetation is primarily native and nonweedy, and the funds are being managed effectively for the long term. In addition, they provide assurances to the regulatory agencies that the conservation goals of the bank are being achieved.

Deciding who should be responsible for conducting this monitoring is one of the many unresolved issues in the dialogue on mitigation banking. Self-monitoring, accompanied by scheduled reports, is the most common approach. Obviously, for private entrepreneurial banks or banks secured by a substantial performance bond slated to be released after certain criteria have been met, there may be a conflict of interest. Methods for reducing this include certification of bank monitors; submission of the name and affiliation of those responsible for the monitoring on all reports and the inclusion of that individual as having shared liability should there be any data found falsified in any way; and, the right, provided for in the implementation agreement, for the regulatory authority to have unrestricted access to the site for monitoring and inspection of the site and all data.

An alternative to self-monitoring is found in New York State's approach to monitoring private-sector hazardous-waste treatment, storage, and disposal facilities. In this instance, the operator of the facility is required, as part of its permit, to provide funding to the state to hire an on-site inspector who is under the employ and supervision of the state. For large complex banks this would be economically feasible. For smaller banks, a designated number of inspections per year by the state or federal government could be paid for by the banker based on a rate determined by the agency. There are administrative and legal constraints that would need to be overcome should this approach be considered.

Monitoring of mitigation banks promises to have some advantages over the present on-site wetland mitigation scenario. At the very least, because mitigation can be clustered and many different development impacts can be compensated for in one place, there will be fewer places to monitor, fewer reports to read, and fewer individuals with whom to interact. In this way, state and federal agencies will probably realize more efficiencies overseeing compliance as compared to the existing mitigation process.

Notes

1. Composite list of data presented in the Tiburon Evaluation of Conservancy Projects (July 1993), USFWS Region 1's Evaluation of Selected Wetland Creation Projects (May 1994), and publication titled: Wetland Creation and Restoration: Status of the Science, Kentula and Kusler, Washington, DC: Island Press, 1990.
2. Shabman, Leonard, Dennis King, and Paul Scodari. *Making Wetlands Mitigation Banking Work: The Credit Market Alternative,* Department of Agriculture and Applied Economics, Virginia Tech, 1993.

3. Willard, Daniel E., and John E. Klarquist. (1992). "Ecological Basis for Watershed Approaches to Wetlands and Mitigation Banks or Cooperative Ventures." School of Public Affairs, Indiana University, Bloomington, Indiana.
4. King, Dennis, "The Economics of Ecological Restoration," 1992.

Conclusion

Lindell L. Marsh, Douglas R. Porter, and David A. Salvesen

The premise of this book, supported by the observations of many of its authors, is that mitigation banking, and entrepreneurial banks in particular, provide a useful option in certain circumstances for mitigating adverse impacts of development in wetlands and wildlife habitats. Various chapter authors have pointed to the ways in which mitigation banking can expedite the permit process, assure compensation for wetland and habitat losses, increase the likelihood of successful mitigation, and provide incentives for private-sector investments in restoration and creation of wetlands and habitats. John Studt and Robert Sokolove in chapter 3 and Leonard Shabman, Paul Scodari, and Dennis King in chapter 6 note that the concept of mitigation banking is similar in its basic aspects to other compensatory mechanisms, such as transferable development rights, air quality credits, and "in-lieu" fee payments by developers. In addition, the primary regulatory agency, the U.S. Army Corps of Engineers (Corps), has endorsed the concept, and even the Environmental Protection Agency (EPA) is warming to the idea, particularly as part of its current interest in watershed planning and ecosystem management. Mitigation banks promise to realize environmental goals while allowing landowners to meet their development objectives unhampered by crippling restrictions.

Nevertheless, after a quarter-century of exploring the concept of mitigation banking, only a few entrepreneurial banks have been initiated; it is difficult to identify even one operating entrepreneurial bank whose track record has won universal praise. Clearly, many environmentalists still view banks with great caution and concern. Jan Goldman-Carter and Grady McCallie's chapter on the environmentalists' perspective points out the high risks attendant to restoring or establishing workable wetlands and habitats. They also suspect that developers will be encouraged to opt for writing a check for mitigation credits rather than avoid filling wetlands in the first place.

Other chapters have pointed to the myriad of difficulties posed in putting the concept into successful practice. Entrepreneurial banks pose grave risks in proving effective, raising questions about the liability of developers who buy credits in banks that ultimately fail. As Leonard Shabman, Paul Scodari, and Dennis King observe in chapter 6, the apparently cheaper use of on-site mitigation, despite the likelihood of failure, puts banks, which often must prove successful before selling credits, at a competitive disadvantage. By contrast, single-user banks and cooperative mitigation arrangements seem simple and straightforward in concept and have proven their effectiveness in satisfying the recurring mitigation needs of utilities, oil production companies, and road builders. Further, the list of types of single-users is expanding into the timber and mining industries, generally in connection with large-scale resource development and conservation plans.

The financial burdens and risks attendant to establishing and operating banks are described with great clarity in chapter 9. Those burdens and risks are augmented by the regulatory context within which banks must function. The first chapter pictures the reluctance of responsible federal and state agencies to embrace the banking concept, justified by a host of reasonable and some unreasonable concerns. The newness of the concept and the lack of documented, successful experience with entrepreneurial banks and, for that matter, with the entire notion of marketing credits have had a chilling effect on agency interest.

There is reason to believe that agency reluctance is diminishing. Some precedent-setting successes have occurred and agencies have demonstrated a growing interest in collaborating with private and public entities to achieve conservation objectives. They are beginning to understand the down side of project-by-project mitigation and the increasingly adverse reaction in the industry to traditional command-and-control regulatory approaches. The greater sensitivity to and knowledge of ecosystem approaches by both public and private sectors, and the expansion of planning efforts for ecosystem management in many areas, provide opportunities for greater long-term assurance for mitigation projects. Experience with broad-scale conservation planning efforts such as special area management plans (SAMPS), habitat conservation plans, and watershed planning is growing, and with it, increasing opportunities for mitigation banking that are embedded within those programs, such as in the West Eugene Wetlands Plan, the San Marcos Creek SAMP, the Carlsbad Highlands Conservation Bank, or the Riverside County Habitat Conservation Agency program.

The question threading through the book is whether entrepreneurial banks have a future. The authors believe that they do—that banks have a legitimate role in meeting preservation and development needs, one that will evolve and increase over time and that for the present will be constrained. Ultimately, we may even see realized the early vision of banks freely trading in the marketplace, selling credits to project managers who need an acre or two of credits to improve their project's feasibility and effectively restoring and managing wetlands and habitats. The Georgia and Florida entrepreneurial banks are a beginning, as is the Orange County, California–operated bank.

Two particular areas of concern about mitigation banks must be addressed if the entrepreneurial banking concept is to take hold. One encompasses the mechanisms necessary to assure long-term effectiveness of the bank—establishment of practical, workable monitoring and maintenance programs, performance standards for measuring success, and financial assurances, such as performance bonds. The other concern relates to the credit vehicle—methods for valuing credits, equating requirements for on-site with off-site mitigation, and allowing advance sale of credits.

So where will entrepreneurial banks find their niche? Certainly, with the new attitudes favoring a collaborative approach to planning and mitigation we can expect a brief surge in establishing mitigation banks such as those in the Carlsbad Highlands. The regulatory context is likely to encourage banks sponsored by owners of the lands (particularly public lands) located in urbanizing areas with a high number of cumulative impacts and relatively low-value or widespread wetlands and habitats (such as southern California, Texas, south Florida, the Washington, D.C./Chesapeake Bay Area, and urbanizing areas within Oregon and Washington). Outside such areas, it is more likely that banks will accompany large-scale development efforts within the context of ecosystem planning.

In the long run, however, it is possible that these ad hoc bank arrangements will give way to a different system in which fees are paid to regional- or ecosystem-related public or nonprofit conservancies based on their average cost of restoring, creating, and managing wetlands and habitats. It is not unlikely that conservation plans for developing areas will provide the basis for mitigating impacts on multiple habitats and ecosystems through regionally imposed impact fees allocated without regard to specific development sites, wetlands, and habitats. The regional planning efforts in southern California, and expanding interest in watershed planning approaches, may build an institutional framework for regional revenue production linked to large-scale conservancy efforts. Utilities, for example, might choose to pay fees to a regional conservancy rather than operating individual-user mitigation banks. This direct connection between mitigation needs and providers provides a simpler, and in some ways more effective, mitigation mechanism.

In imagining the mitigation banking landscape 10 to 15 years from now, we would not be surprised to find

- regional/ecosystem public-sector or nonprofit conservancies accepting mitigation fees that will in turn satisfy regulatory mitigation requirements, with the fees based on the conservancy's average cost of acquiring, restoring, enhancing, and preserving the resource. This probably will occur in the context of an ecosystem or regional conservation plan.
- single-user (or group-user) banks throughout the nation for developers of facilities for utilities, roads, water, and sewers and for resource companies (e.g., timber, mining, and agriculture), probably in the context of large-scale resource development/resource conservation plans.
- scattered entrepreneurial banks operated by, or in joint venture with, the landowner (particularly public-sector landowners) and in many cases located outside of the service areas of the large-scale conservancies.

How does this evolution take place? Assuming that mitigation banking is a good idea, how can the evolution of banks be encouraged? What is needed?

In answering these questions it is helpful to return to the basic function of a mitigation bank, which is to compensate for adverse impacts to natural resources by providing for the conservation of a similar resource in another location. For mitigation banks to provide such compensation, efficiently and effectively, there must be:

Better information Improved information bases and plans, reflecting a better scientific understanding of ecosystems, will help to identify where mitigation banks should be located and when banks would provide the best form of compensation. The expanded use of computer-based geographical information systems could help facilitate the development of such information bases.

Better planning Better planning will provide clearer expectations about future arrangements of both development and conservation. Regional conservation plans such as those being developed in connection with the State of California's NCCP program or south Florida's plan for a sustainable region meet this need, as do the large-scale timber management plans being developed by timber companies in the northwest and the southwest.

Better forms of assurances and trust Assurances begin with trust and are supplemented by the covenants and legal assurances built into banking and other agreements. The character of the bank operator, particularly the long-term caretaker, is important and is enhanced by a history of successful operations. In addition, a basis of trust can be provided by principles and values that pervade the agencies, community, and actors involved in these banking arrangements. These principles and values will be supported by efforts that successfully meet the needs of the various participants. These efforts include specific legal assurances confirming both the nature of the credits for the benefit of the banker as well as the successful completion and operation of the bank for the benefit of the agencies, conservationists, and the environment. We are confident that the specific institutions, legal forms, and broader community principles and values will evolve as specific mitigation banking efforts proceed. In essence, success will breed further innovation and success.

Case Studies

Mitigation banks come in all different shapes, sizes, and institutional arrangements: single-user banks, public agency banks, entrepreneurial banks, and so on. The primary focus of this book has been on the so-called entrepreneurial banks, where mitigation credits are sold on the open market to compensate for wetland losses stemming from development. Two such banks are described here: the Millhaven Bank (case study 1) and the Florida Wetlandsbank (case study 2).

Similarly, Habitat Conservation Plans (HCPs), under the Endangered Species Act, could be considered a form of mitigation banking, in that critical habitat of an endangered or threatened species is protected and preserved in perpetuity as mitigation for development in the species habitat elsewhere. Unlike most mitigation banks, however, where wetlands are created, restored, or enhanced, HCPs typically involve the preservation of habitat only. Two HCPs are shown here: the Coachella Valley Fringe-Toed Lizard HCP (case study 3) and the Riverside County Habitat Conservation Plan (case study 4).

The Coachella Valley Fringe-Toed Lizard HCP focuses on a collaborative effort of a number of local agencies to mitigate the regional development impacts of a single species through the acquisition and maintenance of a preserve. A mitigation fee (impact fee) is one of the funding sources. This HCP underscores the need to address implementation in detail in the implementation agreement. The Riverside County HCP focuses on a very ambitious regional multiple-agency collaborative planning effort to establish a series of conserved habitats as mitigation for development impacts regionwide. Like the Coachella Valley HCP, Riverside utilizes a development impact fee assessed on all development within the historic range of the species of concern. While the conservation plan focuses on a single species, the case study indicates that the plan is providing a framework for the establishment of a multispecies conservation effort as well as individual mitigation banks focused on wetlands and riparian habitat.

An increasing number of large-scale plans, encompassing entire watersheds or ecosystems, are being developed in which mitigation banks are an integral part of the plan. These plans provide a basis to determine the impacts and needs of an ecosystem or species and to establish one or more agencies or organizations that are delegated the responsibility to conserve the wildlife and wetlands resources involved.

Such institutional arrangements can enable a conservation agency to readily determine the monetary cost of conserving the resources involved on a "per acre" or other unit cost basis. For example, an agency or a conservancy can keep track of the average cost of acquiring or creating an acre of wetlands or habitat. As mitigation for its impacts ("debits"), the developer can then be required to pay an amount based on the actual cost incurred by the conservancy, or some proportionately greater sum (e.g., 150 percent), as compensatory mitigation. Two such plans are presented here, the West Eugene Wetlands Plan (case study 5) and the San Marcos Creek Special Area Management Plan (case study 6).

The West Eugene plan fits within the context of wetlands preservation, urban development, and a landscape approach to wetlands restoration. Mitigation and restoration are viewed as key elements in reversing the trend of local wetland losses. The bank is viewed as an institution and a major means of financing the restoration efforts, as well as compensating for wetland fill activities allowed in the plan.

The San Marcos Creek plan focuses on the use of a Special Area Management Plan (SAMP) by a local agency to provide for the reconciliation of the cumulative impacts of development and the conservation of wetlands and other habitat within a 7 1/2-mile segment of the watershed of an urban creek. The SAMP establishes in advance impacts to be permitted, mitigation for those impacts (in the form of one more mitigation areas), and a mechanism to allocate and assess the costs.

Finally, single-user banks, in which a utility, large developer/landowner, or even a timber company will establish a bank from which it will withdraw credits in the future. The Fina LaTerre bank in Louisiana was the first example of this type of bank. Similar banks have been developed to preserve wetlands and wildlife habitat. Two such banks are shown here: the Small Area Mitigation Site adjacent to the San Joaquin

Marsh in Irvine, California (case study 7), and the Aliso Creek Wildlife Habitat Enhancement Project (case study 8). Both banks were created by a major private land developer in cooperation with state and local agencies to offset impacts to riparian areas in southern California. And both projects will involve long-term stewardship by a qualified public agency.

Case Study 1

Millhaven Mitigation Bank

W. Brooks Stillwell

Background

In 1990, Wetlands Environmental Technology, or W.E.T., inc., was formed to construct the nation's first multiuser commercial wetland mitigation bank on a 350-acre site at Millhaven Plantation in Screven County, Georgia. Millhaven comprises a 25,000-acre farm and timber property on Briar Creek, one of the principal tributaries of the Savannah River. The site was once an integral part of the Briar Creek Swamp ecosystem but had been ditched and cut-over to establish a pine plantation. The Briar Creek Swamp lies over the principal recharge area for the aquifer that provides most of the drinking water for the city of Savannah, 100 miles to the south.

Permitting

W.E.T., inc., is comprised of a commercial real estate lawyer, an environmental engineer, and several investors. After optioning the right to acquire a wetland conservation easement on the Millhaven site, W.E.T. prepared extensive studies of an initial 100 acre phase of the project. Berger met with representatives of the Savannah District of the Corps of Engineers (Corps) to determine the regulatory requirements for the project. With little guidance from Washington on how to proceed, the Corps requested that W.E.T. apply for a permit under Section 404 of the Clean Water Act.

W.E.T. applied for the permit in 1992. Over the next two years, the application was reviewed by dozens of regulators in numerous agencies throughout the country. Since the Corps had never previously granted a private company the right to sell mitigation credits for use by third-party clients, the application faced intense scrutiny. The review addressed many theoretical questions concerning mitigation banking, including: What assurance of restoration success will the permittee furnish to the federal agencies? When can the permittee sell credits? Will a bond be required? How will a credit be documented? What terms will be included in the easement? Who can enforce the easement?

In addition to these questions, many individual regulators made site-specific comments concerning the application. For example, one agency criticized W.E.T.'s proposal to plug the ditches on the site instead of filling them completely. Ultimately, these issues were resolved by difficult and protracted negotiations. W.E.T.'s position on the ditches was later vindicated when the bank site suffered from a drought and thousands of amphibians survived in the small pools of water that were retained in the old ditches.

Finally, on December 18, 1992, the Corps issued W.E.T.'s Section 404 permit. The permit allows "the placement of fill into existing man-made drainage ditches for the purpose of restoring or enhancing the natural wetland hydrology to approximately 350 acres of previously drained wetlands. These restored and/or enhanced wetlands can then be used as off-site compensatory mitigation from the established wetland mitigation bank."

To insure the ecological success of the project, the permit includes the following restrictions:

1. W.E.T. must construct the bank and file monitoring reports proving its success before it may sell any credits. The permit requires proof to the Corps of specific mixes of plant species and rates of survivability per acre.
2. After constructing a wetland restoration and enhancement project and filing the required reports, W.E.T. may request that the Corps inspect the project. The Corps must then make a "Preliminary Determination" of the number of acres that it determines have been restored or enhanced. W.E.T. may then sell credits equal to half of those acres.
3. At the time of sale of a credit, W.E.T. must post a bond of $5,000 per credit-acre sold. It must also record a plan of the project site, and a wetland conservation easement on the entire phase of the project, dedicating it as a wetland in perpetuity.
4. W.E.T. must monitor the water level in test wells and the plant survival rate in test plots throughout the project for at least five years. W.E.T. must also repair any damage to the site.
5. After W.E.T. proves to the satisfaction of the Corps that it has restored the predrained hydrology of the site, it may request that the federal agencies again inspect the project. At that time, the Corps, with input from the other agencies, may make a "Final Determination" of the number of acres that have been restored or enhanced. If the Corps makes this determination, W.E.T. can sell credits equal to all of the acres, and the Corps will substantially reduce the bond.

Implementation

In early 1993, W.E.T. proceeded to restore 100 acres under Phase 1 of the mitigation bank. It plugged ditches, altered the topography, and planted over 22,000 hardwood saplings on the site. Its monitoring reports showed a dramatic rise in the water table on the site. On March 14, 1994, the Corps made a "Preliminary Determination" that 60 acres of the site had been "successfully restored to predrained hydrologic conditions." The Corps also modified the permit to reduce the required bond to $1,000 per acre, and allowed a commercial letter of credit to satisfy this requirement since traditional bonding was not available for a wetland mitigation project.

To date, W.E.T. has sold 6.2 acres of credits. The most recent sale was for a price of $15,000 per credit acre. W.E.T.'s environmental consultants have completed initial plans for the development of Phase 2 of the project.

Thus far, the project appears to be successful ecologically. It restored the hydrology of the historic swamp, wetland plant species now thrive, and thousands of birds, amphibians, and reptiles inhabit the site.

Economically, however, the project is not yet profitable. Sales demand has not met projections, preventing W.E.T. from proceeding with the second phase.

W.E.T. attributes initial slow sales to several regulatory factors, all of which are now being addressed by the federal agencies:

1. Most permits for on-site (i.e., not from a mitigation bank) projects contain few, if any, performance standards. They often do not require specific construction plans, monitoring reports, or other safeguards. They face very little regulatory scrutiny, and virtually no postpermit monitoring. Thus, it is often cheaper and less burdensome for a developer to construct a "do it yourself," and ultimately unsuccessful, on-site wetland than to purchase credits from an established, restored, monitored, and successful bank.

2. Some regulators continue to insist that potential bank clients use on-site mitigation, without regard to whether the on-site mitigation is ecologically as beneficial as the wetland restoration available through the Millhaven Bank. Some regulators and engineers believe that on-site mitigation is an absolute regulatory requirement, unless scientifically impossible.

3. Some regulators oppose using mitigation banks for any purpose, because they personally oppose any development that affects wetlands. They believe that using banks will facilitate development. Although this prejudice against mitigation banks contradicts the Clinton administration policy and regulatory guidance issued by the Corps and the Environmental Protection Agency (EPA) it is difficult to overcome. One regulator told a potential major client of the Millhaven Bank that the client could not use the bank for mitigation on projects requiring a Section 404 permit and that the bank could be utilized only in connection with nationwide permits (i.e., for impacts of less than ten acres). Even though this position is contrary to Corps and EPA policy, applicants for Section 404 permits wish to avoid controversy and feel they must acquiesce to

the personal preferences of the individual regulators with whom they must deal on a regular basis.

4. Traditionally, the Savannah District of the Corps has not required mitigation for most wetland impacts that are eligible for nationwide permits. This policy evolved when no mitigation bank was available, when mitigation projects of under ten acres were viewed as unreasonably expensive to developers, and when the Corps lacked the personnel to monitor small mitigation projects.

Notwithstanding these problems, W.E.T. remains optimistic that the Millhaven Bank will succeed economically, as well as ecologically. Several recent developments support this optimism:

1. Both the Clinton administration and the Republican Congress support the increased use of mitigation banking. Proposed amendments to the Clean Water Act support using banks. In addition, the joint federal mitigation banking guidance endorses mitigation banking. This support is slowly percolating down to the regulators in the field.

2. Many wetland scientists now recognize that off-site mitigation is often ecologically superior to on-site mitigation, particularly when the off-site project restores and enhances existing, degraded wetlands (such as the Millhaven bank). The federal guidance specifically recognizes that permittees can use a bank instead of on-site mitigation when the bank is environmentally superior to on-site mitigation.

3. The Savannah District of the Corps recently published Standard Operating Procedures for using mitigation under nationwide permits. In the future, mitigation will be required on most larger projects. In addition, the amount of mitigation required will be influenced by the maturity of the proposed mitigation project. Thus, a client should receive more credit for using an existing mitigation bank, with proven success, than he would receive for using a proposed mitigation project, which poses a higher risk of ecological failure.

W.E.T. is proceeding to develop mitigation banks in other locations. It recently optioned a 500-acre site near Atlanta, on the grounds of a monastery.

Case Study 2

Florida Wetlandsbank

David A. Salvesen

In Florida, about 20 miles southwest of Fort Lauderdale, a private mitigation bank is taking shape on public lands. The 350-acre site, owned by the city of Pembroke Pines, is choked with exotic plants, such as melaleuca trees, that have severely limited its value to native wildlife. The bank, a partnership called the Florida Wetlandsbank, is removing exotic species, enhancing hydrology, and planting native species to create a multihabitat ecosystem typical of the Everglades, including cypress stands, emergent marshes, sawgrass prairie, wading bird–feeding areas, and tree islands. In return, the city will receive an improved wetlands/park with nature trails and picnic areas.

Since opening about a year ago, Florida Wetlandsbank has completed transactions with both commercial and residential developers, including two transferrals totaling 90 acres for a residential development, Spring Valley, in Pembroke Pines; a 1.5-acre transfer for a shopping center in Pembroke Pines; and a 4.5-acre transfer for Sawgrass Preserve, a residential community in Sunrise. The bank is 100 percent privately funded by the sale of mitigation credits.*

Florida Wetlandsbank is required to meet strict criteria for success related to the removal of exotic species, working of the system's hydrology, and recoupment of desirable wetland species. Monitoring—by a certified engineer—continues quarterly for five years after the wetland site is planted. The bank pays the city $1,000 per acre for an escrow fund to maintain the site, plus another $8,800 for a performance bond to guarantee that the work is completed satisfactorily. The performance bond, required by the U.S. Army Corps of Engineers (Corps), will be released when the bank meets the Corps criteria for success. Credits sell for about $40,000 each.

The bank suffered through some of the same permitting delays and dilemmas as developers, namely overlapping jurisdictions and duplication of permitting. For example, after receiving a permit from the Corps in 1993, the bank spent a year and a half completing permitting at the state and local levels. The partnership had to obtain three different permits for the same site and had to meet three different requirements for calculating transfer ratios and bonding.

*Lautin, Lew, "Pioneering Wetlands Mitigation Bank," *Urban Land,* June 1995, p. 12.

Case Study 3

The Coachella Valley Fringe-Toed Lizard

Cameron Barrows

The Coachella Valley fringe-toed lizard (*Uma inornata*), a mere ten inches long, has adapted superbly to life in the loose, wind-blown sand, called blowsand, found in the desert near Palm Springs, California. Its fringed toes allow the lizard to skate deftly along the loose sand while its wedge-shaped snout enables the striped, scaly creature to disappear quickly beneath the sand to avoid predators. Moreover, when it is diving, a loose flap of skin securely covers the lizard's ears to keep sand out.

The diminutive reptile, unfortunately, suffers the grave misfortune of living in the very same areas coveted by developers, with whom it cannot compete. Its habitat dwindled dramatically during the 1970s, and the lizard was endangered of becoming extinct.

In 1980, after years of habitat loss, the fringe-toed lizard was listed as a threatened species by the federal government and an endangered species by the state of California. The lizard was one of several species of animals and plants restricted to a roughly 200-square-mile sand dune system in the Coachella Valley of southern California. Once common, the lizard and its habitat declined wherever wind breaks, country clubs, and buildings blocked sand movement and stabilized the dunes. This listing precipitated concerns and threats ranging from shutting down future economic development in the Coachella Valley to attacks on the Endangered Species Act. Fortunately, cooler heads prevailed. Seeking solutions rather than confrontations, a cooperative process was initiated that culminated in 1986 with the signing of the Coachella Valley fringe-toed lizard Habitat Conservation Plan (HCP) by the U.S. Fish and Wildlife Service (USFWS) as well as by Riverside County and all of the nine affected cities within the Coachella Valley. The HCP detailed aspects of how a preserve system would be created and, importantly, how it would be funded.

Based on 1986 dollars and land values, the HCP estimated that the cost of purchasing the 13,000 acres of private lands needed to establish the preserve system would be $25 million. This sum would also fund an endowment to cover the costs of managing the preserve. While there was complete agreement on the need to establish a preserve system, early on there was substantial conflict as to just who would be responsible for paying the bill for implementation. The private economic development sector argued that due to the federal and state legal nexus (endangered species acts on the federal and state levels), the government should pay for implementation. On the other side, government representatives and private environmental activists countered that the lizard and associated species were threatened by previous and continuing development, and so the developers should pay the bill. Some went so far as to suggest that there should be a one-to-one, acre-for-acre replacement as a means to fund the preserve system's establishment. Under this scenario, a developer would need to purchase an area within a designated preserve equal to the area of the land to be developed. At a price tag of land valued at $2,000 to $20,000 per acre, the developers were not willing to support that alternative.

The inability to achieve consensus on an equitable funding mechanism threatened to dissolve the cooperation that had prevailed up to this step. Finally, a scheme was agreed to in which all affected parties would contribute to the HCP's implementation. The Nature Conservancy (TNC), a nonprofit conservation group, was able to raise $2 million, primarily from residents of the Coachella Valley. Seeing the public interest and support, the federal government, through the Land and Water Conservation Fund, contributed nearly $10 million. The Bureau of Land Management added an additional $6 million to the federal contribution through land exchanges. The State of California's Wildlife Conservation Board added $1 million to the funding mix. This left the implementation of the preserve system about $7 million short of the necessary $25 million. That balance was to be contributed by the developers through mitigation fees paid when they sought grading permits. That fee level was set at $600 per acre; substantially less than acre-for-acre replacement.

That was the funding plan, in rough outline, as it appeared in the HCP. Now, ten years later, did it work? The answer is yes and no, depending on how coarse or fine you look at it. Developer mitigation fees have totaled nearly $5.5 million as of early 1995, all but about 40 acres of private land have been purchased within the largest of the three preserves created under the HCP, and the management endowment is about half of what it will need to be to support annual management costs. Once sufficient fees are acquired to complete the purchase of the largest preserve, fees can then be used to acquire the several hundred acres needed to complete the smaller sites. This sequence in land acquisition was dictated in the HCP. On the surface these numbers suggest that the fee collection and protection activities are in general accordance with the HCP. However, in the years since the signing of the HCP, some concerns have become apparent.

The Coachella Valley fringe-toed lizard HCP was one of the first

HCPs written and implemented, so a few ambiguities and problems in interpretation are inevitable. One glaring problem was the exemption of public works projects from obligation to pay mitigation fees. The goal in this exemption was not to encumber the building of such things as schools and libraries with additional costs such as mitigation fees. In some cases this was interpreted to include municipal golf courses, which converted hundreds of acres of dune habitat with no compensation to the development of the preserve system. This was clearly not the intention of the original crafters of the HCP.

One issue that was impossible to foresee in the preparation of the HCP was the general economic depression that hit California in the late 1980s and early 1990s. With less development, fewer mitigation fees were collected. The result was that less land was acquired within the preserves, and the management endowment was added to at a slower rate than anticipated.

Despite the economic slowdown, one large housing development did get built adjacent to the preserve site, affecting the appraisals for land still needed to be added to conservation ownership. Instead of paying \$3,000–\$8,000 per acre, some parcels doubled their appraised value. The HCP did not include inflation formulas that could be applied to the mitigation fees as land values increased. Fewer dollars now bought much fewer acres. Landowners had higher expectations of value for their land, but fewer mitigation dollars were coming in to pay them.

In 1994, the USFWS contracted for an external audit of mitigation fee collection process. The results of that audit revealed additional issues of concern. For developments occurring within the city limits of the nine cities of the valley, developers paid mitigation fees to the cities or county that then transferred the funds to TNC. TNC was designated in the HCP as the holder of mitigation fees for the various agencies involved in the implementation of the HCP. The external audit revealed the following issues:

- The cities and county had inadequate and infrequent reporting of mitigation fee collections, making tracking of those collections impossible.
- Some cities held onto the mitigation fees for extended periods rather than passing them on to TNC. This resulted in a loss of interest funds for the preserve that would have been gained through TNC's investments.
- There was an uneven interpretation as to what lands were subject to mitigation fees by the cities. This was due to ambiguities in the HCP.
- Nine years passed between the initiation of mitigation fee collection and the first audit. No one conducted annual reviews and evaluations of mitigation reports from the cities.

In the time between the signing of the HCP and the external audit, there was a considerable turnover in the staff of the city and county agencies responsible for insuring mitigation fee collection. With that turnover and with the simple passage of time, there was undoubtedly a loss of understanding by the city and county staff for the criteria of fee collection, the disposition of those funds, and how to report on those funds. With that in mind, the issues raised in the audit are not surprising. The USFWS has the authority and is responsible to insure that the mitigation fees are collected and transmitted appropriately. Lack of staff and other priorities on the part of the USFWS resulted in no one checking with the cities or county to inform them if the collection procedures were correct.

Efforts are underway now, on the part of the USFWS, to rectify the deficiencies in the way the mitigation fees are collected and monitored. It is difficult to estimate the loss of revenue to the preserve system by the cumulative effects of the short-comings of the writing and implementation of this HCP, but they are not insignificant. Based on current and increasing land values, without some modifications there may be insufficient funds to complete the preserve system as it was envisioned by the designers of the HCP. Compounding this is the recent scientific finding that to ensure long-term protection of a vital sand source, the boundary of one of the preserve sites may need to be expanded.

There are many positive aspects that bear mention before a description of the Coachella Valley fringe-toed lizard HCP would be complete. Economic development continues to prosper in the Coachella Valley; no development proposals have been curtailed by the occurrence of fringe-toed lizards. Nearly 20,000 acres of native desert have been protected in a preserve system. Tens of thousands of people visit the preserve system each year to hike, picnic, and enjoy solitude; many receive interesting information about desert natural history from a dedicated cadre of docents. The riparian and palm oasis habitats within the preserve have been revitalized through the complete removal of the otherwise ubiquitous alien shrub, salt cedar. Abandoned agricultural lands are being restored to desert habitats. The cooperative nature of the preserve system's management, including the Bureau of Land Management, USFWS, California state parks, California Department of Fish and Game, and TNC, is certainly a model for effective and efficient land management. On many fronts this HCP has been a success for the affected parties.

The bottom line for the success of this conservation effort is the protection of economic interests of landowners coupled with the long-term viability of the fringe-toed lizard and the dune ecosystem on which it depends. Current lizard populations within the preserve system appear to fluctuate naturally with changing resource availability. So far so good. However, scientists have identified a problem with the preserve design at one site; a vital sand source to the largest dune system has been inadequately protected. Without rectifying that need, many of the existing dunes will eventually migrate off the preserve, with no replenishing sand flowing in to maintain the habitat.

While this HCP has provided a model for cooperative conservation solutions, the jury is still out for its ultimate success. In addition to resolving boundary configurations to complete the habitat protection, the likely shortfall in mitigation income will need to be addressed. Entering into an HCP carries with it risks, the most severe of which is the extinction of the species we are trying to protect. The best assurance against that outcome is continued monitoring, evaluation, and, if necessary, “tweaking” of the process or design.

Case Study 4

Riverside County Habitat Conservation Plan

Alex Camacho

Riverside County extends from the eastern edge of the Los Angeles metropolitan area to the Arizona border. About 90 percent of the county's current population of 1.2 million resides in its western section (about one-third of the county). The Bureau of Land Management owns most of the eastern part of the county. According to the U.S. Fish and Wildlife Service (USFWS) surveys in 1988, over 30,000 acres in the western one-third of the county provide habitat for the Stephen's kangaroo rat (SKR), listed as endangered under both the Endangered Species Act (ESA) and the California Endangered Species Act (CESA). The habitat is scattered in small, isolated areas. About two-thirds of the original habitat has been lost to farming and urban development.

Over the past six years, Riverside County and the seven local municipalities within the county have been actively involved in planning and financing endangered species protection programs and have formed the Riverside County Habitat Conservation Agency (RCHCA). The RCHCA has developed a short-term HCP, for which permits have been issued under Section 10(a) of the ESA and under CESA. In addition, the RCHCA has recently formulated a long-term HCP as well.

To meet the requirements specified in ESA and CESA for the incidental take of the SKR, the RCHCA incorporated in the HCP provisions for the minimization, mitigation, and monitoring of the impacts of the incidental taking. Included was the establishment of seven core reserves for the conservation of the SKR and its ecosystem. As of January 1, 1995, about 91 percent of the land in core reserves was publicly owned. The land was obtained by the expenditure of over $22 million by the RCHCA through its land acquisition program. This core reserve system is the primary means for the conservation of the SKR and its habitat. Developers with proposed projects in the Riverside County area are required to pay an environmental mitigation fee of $1,950 per acre on new development to provide for the acquisition and maintenance of conserved habitat.

A multispecies HCP has recently been proposed for the western Riverside County area, which would incorporate several habitats together into a comprehensive HCP, including wetlands. This multispecies HCP has yet to be adopted, however.

Recently, the RCHCA and the Riverside County Regional Park and Open-Space District have been involved in developing mitigation banking agreements for wetlands in western Riverside County including the Mystic Lake and Santa Ana River regions. Although still in draft, these agreements would provide for the issuance of mitigation credits to developers in exchange for ensuring the creation, restoration, enhancement, and maintenance of wetland habitat.

The Mystic Lake agreement centers on the reestablishment of a functioning aquatic system of 135 acres that mitigates for areas anticipated to be adversely affected within the San Jacinto watershed and projects in the Santa Margarita River watershed with no suitable alternative mitigation available. The agreement is anticipated to include USFWS, CDFG, RCHCA, EPA, and the U.S. Army Corps of Engineers (Corps). The bank lands would be acquired by the RCHCA, with title once acquired being given to CDFG for management and maintenance for 20 years. A Mitigation Bank Enhancement Plan has been prepared delineating the method for determining debits and credits. Bank credits will be available for use as land is acquired and either enhancement activities are completed pursuant to the plan, or funding is guaranteed for these activities. Performance standards have been developed for determining credit availability and bank success, as well as for reporting protocols and for a monitoring plan that includes contingencies of success and requires remedial actions to ensure bank success.

The Santa Ana River mitigation bank proposal involves the creation, restoration, and enhancement of 1,500 acres of riparian habitat of the endangered Least Bell's vireo along the Santa Ana River. This would consist of the removal of nonnative vegetation, revegetation, and promotion of natural revegetation to restore the native quality of the habitat. Restoration, monitoring, and maintenance would be undertaken by Riverside County Park for 20 years (though not explicitly mentioned in the agreement, it appears that the land is owned by Riverside County Park). Upon restoring the area, the mitigation project known as the Santa Ana River Habitat Recovery Project would be available to be used for off-site compensatory mitigation for unavoidable wetlands impacts in the Santa Ana River watershed. Compensation credits would be available for such projects in need of compensatory mitigation credits in the Santa Ana River watershed. As a single-party bank, the credits would be used by the county as mitigation for projects such as roadway widening, maintenance activities, road crossings, bank stabilization for erosion or flood prevention, and outfall structures. Funding would be provided for the project by the Riverside County Park. Compliance would be monitored by the Corps, EPA, and USFWS, all proposed signees to the agreement.

Case Study 5

West Eugene Wetlands Bank

Steve Gordon

Background

The city of Eugene (1995 population of 120,000) is the larger city in Oregon's second-largest urban region (Eugene-Springfield 1995 population of 200,000). Eugene is located about 110 miles south of Portland along the Interstate 5 corridor. Nestled at the southern end of the Willamette Valley, western Eugene lies within the Amazon Creek watershed, a tributary of the Long Tom River, which flows into the Willamette River north of the Lane County/Benton County border.

Oregon and the city of Eugene are renowned for their planning history and sophisticated approaches to comprehensive land use planning. Comprehensive zoning of the city of Eugene dates back to 1948. An early long-range land use plan dates to 1959. In 1972, the city adopted its "1990 Plan," complete with an urban services boundary. In response to the 1973 Oregon Land Use Planning Act, the city prepared an update of its comprehensive plan in 1982, and that plan was "acknowledged" as meeting all 15 applicable statewide planning goals by the Oregon Land Conservation and Development Commission in August 1982. The Metropolitan Plan concentrated existing and future urban uses within an urban growth boundary in order to protect agricultural and forest resource lands in surrounding rural areas. Eugene's major industrial area was designated in western Eugene, along Highway 126 and the railroad. Over the years, more than $20 million was invested in public infrastructure (roads, water lines, sanitary sewers, and storm drainage improvements) to serve these uses.

In 1987, over 760 acres of wetlands were inventoried in west Eugene; most of them located on vacant land that was planned, zoned, and serviced for future industrial uses. State and federal wetland laws, however, threatened to curtail future economic development that had been planned carefully for years. An existing electronics firm's plans to expand were jeopardized because the firm's property was surrounded by wetlands. The situation pitted wetlands versus jobs in an area that was just recovering from a deep recession in the early 1980s. A classic battle loomed before the community.

In examining the situation and alternatives, the City Council decided to develop a comprehensive wetland management plan with the following objectives:

1. To use the best information available and affordable to help the community understand the choices available;
2. To find a balance between environmental protection and sound urban development that meets state and federal laws and regulations;
3. To provide opportunities for involvement of all interested segments of the community in plan development; and
4. To turn a perceived "wetlands problem" into a "wetlands opportunity" for the community.

The council concluded that the case-by-case, individual wetland permitting process under Oregon and federal wetland laws would not solve the regional issues in western Eugene.

Developing the Wetland Plan

In 1989, the plan work program and citizen involvement program were approved by the city. The Lane Council of Governments was selected as the project manager. An interdepartmental team, dubbed the "Wetheads," was formed to serve the staff-planning function. The U.S. Environmental Protection Agency (EPA) awarded a $50,000 grant to prepare an advanced wetland inventory and assessment of wetland functions and values. Using the 1989 federal delineation manual, 1,300 acres of wetlands were identified (almost doubling the area of the earlier inventory that was based on vegetative cover and did not inventory disturbed agricultural wetlands). The impact on industrially zoned land was even greater than originally described, worsening community concerns, while national wetland debates raged on in the media.

The proactive citizen involvement program involved a series of workshops, newsletters, field trips, direct mailings, personal interviews, and field visits, as well as other techniques to include owners, environmental groups, elected and appointed officials, and other interested citizens. Key state and federal agencies were involved through a technical advisory committee (TAC) and other local meetings.

Wetland definitions, functions, field identification methods, goals, and alternatives were presented to the public and TAC. People were able to respond to maps, opinion surveys, draft inventories, a vision statement and graphics, criteria for protection and development, and to alternatives ranging from "no build" to "fill 'em all." By August 1992, the plan was adopted by the city of Eugene– and Lane County–elected officials. About 1,000 acres were in the "protection" or "restoration" categories, and about 300 acres were in the "development" category. Most of the "development" wetlands were disturbed by urban or agricultural uses or were small, isolated wetlands that were affected by surrounding urban development. The protection and restoration categories fit into a connected system with the streams linking them.

The multiple objective plan addressed many issues: stream protection and restoration, flood control, stormwater conveyance, water quality, recreation, education, research, urban development and certainty, natural diversity, wildlife habitats, threatened and endangered species, wetland protection measures, financing, wetland system management, mitigation, and a regional wetland mitigation bank.

Various tools were used to implement the plan. Among them were

- implementing an acquisition program (along with conservation and utility easements, land exchanges, and donations);
- drafting a natural resources zoning district with wetland buffers;
- drafting a waterside setback ordinance;
- seeking other funding for planned projects (e.g., U.S. Army Corps of Engineers (Corps)—Water Resources Development Act, Section 1135 funds, for stream enhancement and wildlife habitat improvements estimated at $5 million; Oregon Department of Transportation, Intermodal Surface Transportation Efficiency Act enhancement funds for a two-mile Amazon Creek stream enhancement and bicycle path construction project at $2.4 million);
- forming a partnership (Wetland Executive Team or WET) among city, U.S. Bureau of Land Management (BLM), and The Nature Conservancy (TNC) to jointly manage the system;
- preparing and adopting a Stormwater Management Plan (1993) for the city (the first in the nation for its city size category and awarded a National Pollution Discharge Elimination System (NPDES) permit by the Oregon Department of Environmental Quality in 1994);
- completing environmental assessments on BLM management units so trails, viewing platforms, and other recreational and educational improvements could begin construction in 1995 on federally owned lands;
- drafting a local wetland permitting process to handle the Oregon wetland permitting program;
- drafting a memorandum of agreement with the Corps, EPA, and Oregon Division of State Lands (DSL) to meet new federal guidance for mitigation banks;
- undertaking mitigation banking projects in 1993 (to meet Oregon requirements) with an interim memorandum of agreement drafted in 1995.

The wetlands plan was prepared within the context of federal advanced identification and special area management plan. The plan also conformed to the letter and intent of Oregon's wetland conservation plan statute and administrative rules.

As a refinement to the Metropolitan General Plan, the West Eugene Wetlands Plan goals, policies, and plan map were adopted by ordinance, giving them the effect of local land use law.

The mitigation chapter of the plan contains goals, policies, and recommended actions aimed at achieving the following objectives:

- Avoid and minimize impacts to all wetland sites that meet the plan's protection criteria.
- Where impact is unavoidable, compensate for losses commensurate with the level of impact giving priority to establishing the basic physical wetland parameters (water, topography, connectedness) that eventually results in full functioning and diverse wetland habitats.
- Where full-functioning wetlands are to be impacted, in-kind replacement of significant functions and values will be required. The overall mitigation program, however, will be guided by the ecological characteristics of the regional landscape and not necessarily by specific case-by-case impacts. For disturbed agricultural wetlands, mitigation requirements will be determined using historic wetland types presumed to have existed prior to disturbance and the desired mix by the public. Incentives will be provided to mitigate in advance of impact in the form of replacement ratios that are less than the ratios for compensating at the time of impact.
- To increase the certainty of success and to achieve the goal of a connected system of wetlands and waterways, mitigation efforts are targeted for areas that once exhibited, or currently exhibit, proper wetland soils and moisture conditions. The primary mitigation sites are "disturbed agricultural wetlands." These sites are missing at least one of the three wetland parameters (water, hydric soil, wetland vegetation) and due to nonwetland activities (agricultural uses) occurring on these sites, their existing wetland value is relatively low. As such, for mitigation credit purposes, the plan gives more credit for the enhancement of these sites than for other "low" quality wetlands that exhibit all three wetland parameters.

The primary means to achieve the plan's mitigation goals is through the establishment of a regional wetland mitigation bank. These are areas where the most suitable lands for mitigation are identified, acquired, designed, constructed, monitored, and managed in advance of wetland impact, and incentives are provided that encourage the use of the bank by those seeking a wetland impact permit. Because the plan will enhance and restore more wetland acreage, functions, and values than will be lost to development, the excess capacity will be available for mitigation credit to properties located outside the west Eugene study area and within the urban growth boundary, as well as to the Eugene Airport proper.

West Eugene Mitigation Bank Program

The city of Eugene, through its Public Works Department, will operate the mitigation bank with intergovernmental agreements or memoranda of agreement among other key agencies. The term "bank" represents the institution that will combine financing, permitting, mitigation, monitoring, managing, and administration of the bank. Bank fees will be placed in a special city fund for accounting purposes, with funds dedicated to bank activities. The TAC will serve as the mitigation review board and will advise the Corps and DSL. The Wetland Execution Team (WET) will serve to provide direction on the management aspects of the bank that involve other local partners (i.e., BLM and TNC). The Wetheads will advise the Public Works Department on mitigation site selection and design. DSL and the Corps have final approval of mitigation plans and certification of credits in the bank. Annual reports on permitting activities, restoration, sale of bank credits and monitoring activities will be presented to the TAC.

The city intends to monitor the cost of mitigation and alter the fee for credit purchase annually, based on the previous year's experience. The Eastern Gateway Wetland Restoration project (see later), the first local attempt to implement the mitigation banking concept, cost about $30,000 per acre, but this project is thought to be on the high side due to the high costs of removing old fill material from the site.

Studies of eight large mitigation sites have been completed and the sites have high probability of success for restoring lost wetland acres and functions and values. The soils are hydric, and there appears to be enough annual precipitation and surface runoff on the sites to presume that hydrologic factors can be met. Much of the site preparation involves removing agricultural crops and nonnative plants. Weeding nonnative plants during the early years of mitigation will become a major activity undertaken to allow seeded native plants an opportunity to become well established.

The Eastern Gateway project was established to mitigate wetland losses caused by four development projects: (1) expansion of the Eugene Airport runway, (2) construction of low-income housing, (3) construction of a warehouse and parking lot, and (4) improvement of an electric utility substation. Together, the four projects resulted in a combined wetland loss of approximately 1.7 acres. To offset this loss, the DSL required the restoration of about five acres of wetlands (exchange ratios of 3:1 for three projects and 3:1 for the remaining project).

The 17-acre site selected for the creation of a mitigation bank was purchased by BLM in 1993 using federal Land and Water Conservation funds. It contained approximately nine acres of wetlands recommended for protection under the West Eugene Wetlands Plan. About 7.3 acres were slated for restoration. The goal of the restoration was to reestablish the historic wetland types that naturally occurred on the site. This involved removing about 20,000 cubic yards of fill material and creating a wet prairie, and emergent wetland, as well as enlarging an existing pond on the site. Most of the fill was relatively shallow, between one and two feet deep, but some portions of the site were buried with over two feet of fill. The project was designed, constructed, and planted in 1993. The restored wetland encompassed 7.3 acres; 5 to mitigate for the loss of 1.7 acres plus 2.3 acres to be banked for future use.

Accomplishments

In addition to creating the first mitigation bank, with others to follow, other noteworthy achievements of the West Eugene Wetlands Plan program are summarized here:

Partnerships

The city, BLM, and TNC formed a formal partnership in February 1994 to jointly manage the wetland system. This partnership, known as the Wetland Executive Team (WET) recently signed a memorandum of understanding that assigns responsibilities to the three parties. WET is also drafting its first work program, complete with future capital projects. WET has also directed staff to examine land exchanges to better consolidate ownerships for more efficient land management.

Land Acquisition

The city and TNC lobbied Congress successfully for $4.47 million in land and water conservation funds to the Eugene District of BLM to buy wetlands based on the Wetlands Plan. BLM has obligated those funds in cooperation with the other partners with sensitivity to local landowners' wishes. Other agencies have purchased wetlands for protection and restoration purposes. Collectively, about 80 percent of the plan's wetlands in the protection or restoration categories are in public or non-profit group ownership. The ownership is broken down as follows:

Agency	*Acres*
BLM	555
City	204

TNC	325
ODOT	159
Lane Co.	19
TOTAL	1,262

These acre totals include some adjacent uplands that provide wetland buffers and broader representation of upland ecosystems adjacent to the wetland complex.

PROTECTION

Many sites with critical habitat or rare species are in public or TNC ownership. TNC owns 325 acres, including a 60-acre wet prairie site that contains six species of sensitive plants that are on the federal or state threatened and endangered species lists or are candidates for such lists. Recently, the Bonneville Power Administration (BPA) assisted TNC in acquiring 125 acres through the Northwest Power Planning Act's wildlife restoration program. TNC and BLM are also working to protect the habitat of a listed insect, Fender's blue butterfly. Additional research is being conducted on the western pond turtle. TNC and BLM bring biologists and botanists to the program, skills that are lacking in many local government offices. The city and TNC jointly fund a TNC position of plant ecologist who also serves as the TNC steward for several southern Willamette Valley protection sites out of the Eugene public works offices.

The West Eugene Wetlands Plan continues to provide the context for individual decision making and better policy decisions in west Eugene. It has paved the way for federal lobbying efforts, for seeking grants funds, and for drawing new partners to the table. Discussions are underway to add youth corps and the U.S. Army Corps of Engineers as partners along with the city, BLM, and TNC. The vision expressed through the plan has a magnetic effect and stimulates new perspectives and new ideas as implementation takes place. The plan and its staff have received state, regional, and national awards. The plan is touted as a model for resolving urban wetland issues.

The plan was adopted by the city of Eugene and Lane County in 1992. In 1992, it was approved as a postacknowledgment amendment to the Metropolitan Plan by the Oregon Land Conservation and Development Commission (LCDC). In 1993, BLM approved the plan to guide its land acquisition program. In 1994, the Oregon Division of State Lands approved the plan as Oregon's first "wetland conservation plan," paving the way for city administration of the state wetland impact permitting system. In 1994, the U.S. EPA and Corps also approved the plan, paving the way for a more streamlined federal wetland-permitting process in cooperation with the city of Eugene. In all, seven organizations at the local, state, and federal levels have adopted or approved the plan, which speaks highly of the quality of the document and the process used in developing it.

Case Study 6
San Marcos Creek Special Area Management Plan

Alex Camacho

The city of San Marcos is located in the northeastern region of San Diego County between the Interstate 15 and 5 freeways, at the base of the Merriam Mountains. San Marcos Creek flows from these mountains, through the city, and ultimately into an estuary on the Pacific Ocean. The floodplain includes riparian woodland, vernal pools, wet meadows, and freshwater marshes and provides a corridor for upland species. Endangered plant species include the San Diego thorn mint, the Thread-leaved brodiaea, and the San Diego button-celery. Endangered wildlife species include the Willow flycatcher and the Least Bell's vireo. In addition to these endangered flora and fauna species are numerous proposed and candidate species.

The city is rapidly urbanizing with new office and commercial buildings, a state university campus, two regional hospitals, a civic center, and major residential development, all along the creek. In anticipation of future development and to prevent flooding that may occur as a result of development along the floodplain, the city envisioned a "green" corridor that, like the city of San Antonio, Texas, might provide a linear thematic focus running through a new downtown area. As a first step in accomplishing this, the city developed the Special Area Managment Plan (SAMP) to address flood control, as well as the conservation of the creek-related ecosystems. (The city of San Marcos is also working on a Biological Resource Management Plan which seeks to balance new development with biological resource preservation.) Allowed impacts to the creek ecosystems were established and, as mitigation, the city committed to set aside a 57.3-acre compensation bank formally called Twin Oaks Valley Advance Mitigation Area, as well as possibly other areas. These other areas would be acquired and developed as wetlands in advance of impacts to the creek, with the city committing to hold and maintain the areas over the long term and to convey a conservation easement over such areas to the California Department of Fish and Game to assure that the areas will be conserved.

The SAMP allows the city to then allocate the costs of the mitigation areas to individual developments that may occur along the creek using various mechanisms ranging from impact fees to assessments under the new downtown specific plan that is in preparation.

The SAMP provides that the amount of mitigation to be provided may vary. If the mitigation is commenced three years before the development project occurs, a mitigation ration of 1:1 is applied, (i.e., one acre of land may be acquired and mitigated for every acre of land impacted by the development). Otherwise, if the acquisition and mitigation of sites is planned concurrently with the development project and is not a component of the Advance Mitigation Area, the ratio increases to 1.5 mitigation acres required for one acre of impacted wet meadows or two mitigation acres for one acre of impacted riparian environments.

A major frustration of the city in establishing this program has been the relative lack of sophistication and the project-by-project mentality of the state and federal agencies, including the U.S. Army Corps of Engineers, the Environmental Protection Agency (EPA), federal Fish and Wildlife Service, the State Department of Fish and Game, and the California Water Resources Control Agency (under Section 401 of the Clean Water Act). This has been evidenced by a failure of the agencies to seriously commit to the "planning" process, preferring instead to wait until a "formal" permit application was made. While in theory a SAMP allows regulators and devlopers alike to anticipate and address environmental impacts in advance, thus expediting the permitting and mitigation process, the agency's continued reliance on case-by-case review of development proposals has limited the SAMPs' effectiveness.

Case Study 7

San Joaquin Marsh Small Area Mitigation Site

John M. Tettemer

Background

The 17-acre Small Area Mitigation (SAM) site was created in anticipation of impacts to wetlands and riparian woodland areas resulting from 11 land development projects planned by The Irvine Company (TIC). The development projects included roads, flood control projects, research and development facilities, a bioscience center, and commercial and residential areas. Individual impact areas ranged from 0.3 acre to 3.0 acres, averaging 1.5 acres.

It is common for projects with small impacts to develop wetland mitigation sites that typically are surrounded by development that may have limited habitat value because of nearby urban activity. In addition, ongoing maintenance and management are not cost effective and difficult to arrange for such small sites. TIC realized a better solution was needed: a practical, reliable, affordable solution that made a positive contribution to regional habitat needs.

Site Selection

TIC decided to seek a site on which to consolidate all of the adverse impacts to wetlands and wildlife from its proposed development. The search led to the San Joaquin Freshwater Marsh (Marsh). Years ago, the area was a popular duck-hunting site, made up of natural and man-made marshes and duck ponds. Part of the Marsh is now owned and operated by the University of California Natural Reserve System (UCNRS). The largest portion is owned by TIC and has been kept in a natural condition except for areas dedicated to seasonal wildlife ponds.

The site selected offers sufficient acreage to mitigate all of the anticipated impacts of the 11 projects in one integrated project. Beyond the benefit of size, the SAM site is isolated from urban areas and it borders the existing UCNRS preserve. Soils, surface and groundwater hydrology, and topography of the site appeared favorable. Naturally occurring riparian vegetation (Mulefat) and natural seed sources were available to promote plant diversity.

Consultations were held with the U.S. Fish and Wildlife Service (USFWS), U.S. Army Corps of Engineers (Corps), and the California Department of Fish and Game (CDFG) to determine whether the Marsh was acceptable as a consolidated mitigation site for Clean Water Act Section 404 permits and California Resources Code Section 1603 Streambed Alteration Agreements. The resource agencies agreed that the site was satisfactory, provided that plant survival and growth standards were met. The agencies remained closely involved throughout planning, design, planting, and monitoring.

Design and Construction

TIC asked John M. Tettemer & Associates, Ltd. (JMTA), to manage preparation of planting plans; oversee contractor activities during site preparation, planting, and grow-in; monitor tree growth and water conditions; and recommend ongoing operation and maintenance strategies. The presence of a project manager throughout the project offers the advantages of complete knowledge of site conditions and project history and an understanding of what works and what does not. Issues addressed by the project manager include irrigation amounts and schedules, weed removal, use of pesticides and herbicides, site access, irrigation system adjustments and repairs, direction of contractor activities, field monitoring and reporting, groundwater monitoring, liaison with resource agencies, and communications with TIC.

To minimize costs, encourage innovation, and allow flexibility, the contract specified only a planting plan, site preparations, and a water source. Design of the irrigation system was left to the contractor, subject to approval. The contractor was required to maintain the plantings for at least nine months, with an option to extend maintenance for one year. The contractor is responsible for 100 percent survival of the planted trees throughout the life of the contract. Providing wide latitude in terms of design and implementation allows the contractor to propose the most economical approach. On the other hand, careful contract management attention is required to assure that specified results are obtained and problems are dealt with promptly.

The SAM site is intended to mitigate impacts to riparian wetlands; therefore, the plantings consist of willows and cottonwoods, as directed by the USFWS and CDFG. The site was divided into eight planting areas. Each area was planted with Arroyo Willow, Black Willow, and Western Cottonwood, or a combination thereof.

Previous experience at this site and others has demonstrated the importance of site preparation to create a hospitable environment for each tree seedling and assure proper operation of the irrigation system. All old irrigation pipe, debris, and dry brush was removed. The site was disced to aerate the soil, break down old vegetation remnants, and prepare the soil for shaping into windrows.

A tractor with an earth-shaping plow was used to construct planting windrows two feet high and six feet apart. Planting holes for one-gallon plants were dug on the side slopes of the windrows 12 inches to 18 inches above the bottom of the windrow. Placing the holes well up on the side of the windrow assures well-drained soil conditions for early root growth and keeps young roots out of any standing water that might result from rain or broken irrigation lines. Drip lines were placed along the windrows with emitters for each hole. The irrigation system was activated to check the system and presoak the holes.

One-gallon rooted trees had been previously ordered and were available when the holes were ready. Over 13,000 trees were planted—the willows on eight-foot centers and the cottonwoods on ten-foot centers. Planting began in April of 1990 and continued for four months.

An existing operational well at the Marsh preserve provides irrigation water for the SAM site through a four-inch-diameter (PVC) pipe installed as part of the mitigation project.

An irrigation system was installed to sustain the new trees until the root systems reach available groundwater. At the time of planting, groundwater was more than 14 feet below the surface. Soil examination indicated that groundwater had historically been within a few feet of the surface, but a prolonged drought in the late 1980s and early 1990s resulted in a lower water table. Thus, an irrigation system was essential. Three piezometers and six tensiometers were installed to monitor groundwater levels and soil moisture. The information collected was used to determine irrigation needs.

Drip irrigation was chosen as the initial system rather than flood irrigation for better reliability and control, more efficient water application, and to minimize weed growth. In general, these benefits accrued; however, drip systems are subject to some maintenance and operation problems, including plugging of emitters by algae and chewing of the lines by animals.

The irrigation strategy was to deep-irrigate the plants to encourage downward growth of roots toward groundwater. During the first season, water was applied eight hours per day, Monday through Friday, delivering about four gallons per tree per day of irrigation. The tensiometers were monitored closely. As the tree canopies and root systems grew, water requirements increased. In April 1992, the irrigation program was revised to increase deliveries from 20 gallons per tree per week to 44 gallons per tree per week.

The 1991–92 season brought the first year of normal rainfall since 1985–86. In January 1993, the drought ended. The 1992–93 season produced double the mean annual rainfall. Irrigation was suspended in January 1993 and remained off except for brief activations, the last of which was in September 1993. Supplemental water was applied again during the summer of 1994. The irrigation system was modified from drip to flood during 1994. Every second or third furrow can be flooded. This has eliminated emitter maintenance and improved water supplies available for expanding understory growth. Soil moisture, groundwater depth, and tree conditions are being monitored so that irrigation water can be applied, if needed. The trees may be self-sustaining at this point. Continued monitoring will determine whether this is the case.

Management and Maintenance

Pests (galls, leafhoppers, rust, psyllid, mites, aphids, and caterpillars) have not posed a problem. Weeds, however, are a major threat to new trees. They compete aggressively for water and sunlight and will crowd out the trees if allowed to do so. Accurate, effective control is hard to obtain with herbicides. Experience indicates that aggressive hand weeding is the best approach. On this project, the contractor was instructed to maintain a 12-inch weed-free radius around each tree. One of the most troublesome weeds was the Russian thistle. The plant is so large that maintaining a 12-inch weed-free zone meant clearing a substantially larger area. At least two full-time workers were required during the warm months of the first season. As the trees grew larger they were not bothered by the weeds. In many areas the "weeds" are part of a thriving understory complex.

To monitor the health and growth of the trees, 73 trees selected randomly were measured monthly. Reports were submitted to TIC and to the resource agencies. By the end of the second year, the average height was 14.8 feet. The taller trees were 21 feet tall. A year later, in 1993, many trees were over 20 feet tall, and the tallest were over 25 feet tall. Many species of birds and other native wildlife have been seen and heard during informal walk-throughs.

By 1993, the Arroyo Willow planting areas were so fully developed that in some groups the trees were experiencing overcrowding. As a result of the initial spacing (eight feet on centers along the windrows), the trees were cramped, with leaf growth in the upper 40 percent of the tree and little or no understory. This is particularly true of the Arroyo Willows. To encourage plant diversity and understory growth, JMTA developed, and the resource officials approved, a thinning plan for removing some trees in a random, natural-appearing pattern. Thinning occurred in late 1993 and early 1994. The cut trees were chipped for compost, which was hauled away by local landscape companies. By early 1995, understory growth had started and trees within and adjacent to the removal area are sprouting new branches in the lower part of the trees.

A condition of resource agency approval is a guarantee of permanent management of the completed bank. From the outset, it has been the understanding and intent of the parties that the completed site will be dedicated to a public agency for its stewardship.

Case Study 8

Aliso Creek Wildlife Habitat Enhancement Project

John M. Tettemer

Background

Aliso Creek Wildlife Habitat Enhancement Project (ACWHEP) is a riparian wetland mitigation bank developed by the Mission Viejo Company and Orange County to offset impacts of future land development and public works projects constructed under the Corps's Section 404 permits and California Section 1601–1603 Streambed Alteration Agreements.

ACWHEP evolved from work by John M. Tettemer & Associates, Ltd., as it assisted the Mission Viejo Company in developing a mitigation strategy in anticipation of impacts of several planned developments. The effort was driven by one large development and many smaller projects to be built over several years. The scope for ACWHEP was expanded to create a bank that would satisfy all foreseeable Mission Viejo Company mitigation needs.

The site selected offered isolation from urban activity, a water supply from perennial urban runoff and storm flow, and the opportunity to enhance existing patches of riparian vegetation. This site was along Aliso Creek (Creek) in a large open-space area that the Mission Viejo Company had previously dedicated to the county for the Aliso/Wood Canyons Regional Park. The Mission Viejo Company approached the county with a proposal for a joint venture.

As owner of the land, the county was interested in an arrangement that would provide mitigation for future county projects. It was agreed that the county would provide a 70-acre site along the Creek. The Mission Viejo Company would design the entire 70-acre mitigation bank and would receive the use of 35 acres for its own mitigation requirements. The Mission Viejo Company would proceed to develop its 35 acres, including construction of an irrigation headworks on the Creek that would serve the entire 70 acres. After two years of operation, the completed 35 acres and the irrigation headworks would be turned over to the county for perpetual operation and maintenance. The county will develop its portion as needed to meet its project mitigation requirements.

The selected site is located in the historical flood plain of the Aliso Creek. At one time, the Creek probably was an intermittent stream, as is typical in many southern California streams. The streambed itself probably was a shallow, sandy-bottom wash bed, with larger flood flows spread over the valley floor.

The stream is now incised with banks up to ten feet high. Storm flows are contained within the channel. There is an upland area adjacent to the Creek that, if watered, could support riparian vegetation. Sparse patches of Mulefat (*Baccharis glutinosa*) and Mexican Elderberry (*Sambucus mexicana*) remain on the old floodplain or upland area, but the willows, cattails, and other wetland species are generally limited to the channel edges where there is adequate moisture.

Planting and Irrigation

Mission Viejo Company's 35 acres is subdivided into 13 planting areas ranging in size from 0.82 acre to 6.17 acres. The planting palette includes:

Common Name	Scientific Name
Arroyo Willow	*Salix lasiolepis*
Black Willow	*Salix gooddingii*
Red Willow	*Salix laevigata*
Yellow Willow	*Salix lasiandra*
California Oak	*Quercus agrifolia*
Western Cottonwood	*Populus fremontii*
California Sycamore	*Platanus racemosa*
White Alder	*Alnus rhombifolia*
Mexican Elderberry	*Sambucus mexicana*

The palette and planting plan were developed in consultation with the USFWS, CDFG, and the county to achieve a natural clustering effect, leaving room for understory development, and to meet the county's park requirements. Approximately 10,520 trees and shrubs were planted in the initial 35 acres.

Unwanted upland species and low-quality wetland plants were removed by mowing and discing. Existing areas of good quality willows, elderberry, and mulefat were preserved. Irrigation was not directly provided to these areas, as they have demonstrated they are self-sustaining. It is possible that root development in the irrigated areas will reach a point where the new plantings will also become self-sustaining.

The irrigation system consists of the irrigation headworks structure, transmission lines, irrigation lines, furrows, tailwater ditches, downdrains, and flushing lines.

The irrigation headworks structure impounds water in the Creek at an elevation from which it can flow by gravity to all planting areas. Grates and screens remove floating materials before they enter the headworks. A desilting system separates sediments from the irrigation water. The first two cubic feet per second (cfs) of flow are returned to the Creek to maintain natural creek values and functions (including sediment transport) downstream. The desilted water is delivered to two main transmission lines, one on each side of the Creek. Storm flows in excess of pipe capacity pass over the spillway and return to the Creek.

The transmission lines are made of polyvinyl chloride (PVC) pipe and sized to carry five cfs each, sufficient to supply the entire 70-acre project. Pipe size at the headworks is 20 inches, reducing to 12 inches at the downstream end of the initial 35-acre development.

Delivery of water to the trees is accomplished by a gravity-fed system of irrigation lines and furrows. The PVC irrigation lines bring water from the transmission lines to the head of each furrow. Plastic ball valves at the head of each furrow allow precise adjustment of flow. Furrows, which were constructed using a tractor with a blade shaped to create the desired cross-section, normally take three to eight gallons per minute. The trees were planted on the side of the windrow, out of the water-carrying furrow.

To assure that all trees are watered, some water is allowed to reach the end of the furrow. The optimal valve setting is that which delivers water to the last tree with minimal, if any, overflow. Should there be any overflow, it is collected in tailwater ditches. These ditches carry tailwater to downdrains, which return the water to the Creek in a controlled manner to prevent erosion.

Valves and flushing lines were installed to provide the ability to flush out sediment and debris that might accumulate in the irrigation lines. To date, these flushing lines have not been needed.

Multiagency Agreement

The responsibilities of the participants and the definition of mitigation credits are documented in an agreement among the USFWS, CDFG, Orange County, and the Mission Viejo Company. The agreement contains the performance criteria to be used to calculate mitigation credits earned by the Mission Viejo Company and the County. These criteria are as follows:

Time from Planting of Vegetation	Criterion	% of Potential Credits Granted
0 Year	Initiation of headworks structure construction	45

1 Year	80 percent survival of originally planted vegetation	60
2 Years	Average height of planted: woody species = 6 feet percent cover = 40–60 %	70
3 Years or less	Average height of planted: woody species = 10 feet percent cover = 80%	80
4 Years or less	Average height of planted: woody species = 16 feet percent cover = 90%	90
5 Years or less	Average height of planted: woody species = 20 feet percent cover = 100%	100

Performance

The official completion date for the 35-acre planting for purposes of the agreement criteria was September 1993. One year later, the site was monitored and measured. It was determined that the first-year survival rate was 88.8 percent, exceeding the criterion established in the agreement. Replacement of the lost trees is proceeding.

In a small area of the site that had been planted a year earlier as a test area, all measured trees exceeded the two-year height criterion of six feet, with most trees exceeding 12 feet in height. Even the one-year-old trees met the two-year height criterion in most cases.

Canopy is measured using a random sample of trees of each species. Around several trees, the area that mature trees should occupy (100 percent canopy) is staked. Canopy coverage is estimated during the summer by visually estimating the percentage of the staked area covered by midday shade. The two-year-old trees were determined to have canopy coverage of 54 percent, again meeting the two-year criterion.

More than 10,000 young trees were planted at ACWHEP, enhancing the habitat value of the Creek valley floor. Based on initial measurements, it appears that the plants will easily meet the success criteria in the agreement.

Summary

The scattered habitat areas that existed previously supported small numbers of birds and other wildlife. Now the Small Area Mitigation (SAM) site and ACWHEP are attracting new wildlife visitors. Every indication is that these new riparian forests will take their place as major regional wildlife habitat resources. The success of these two sites demonstrates that thoughtful site selection and planting, a reliable water supply, and diligent attention to weeding and irrigation management during the critical early stages as well as the full support of the project proponents can result in the successful creation of wetland mitigation banks.

Appendix

Federal Guidance for the Establishment, Use and Operation of Mitigation Banks

AGENCIES: Corps of Engineers, Department of the Army, DOD; Environmental Protection Agency; Natural Resources Conservation Service, Agriculture; Fish and Wildlife Service, Interior; and National Marine Fisheries Service, National Oceanic and Atmospheric Administration, Commerce.

ACTION: Notice

SUMMARY: The Army Corps of Engineers (Corps), Environmental Protection Agency (EPA), Natural Resources Conservation Service (NRCS), Fish and Wildlife Service (FWS) and National Marine Fisheries Service (NMFS) are issuing final policy guidance regarding the establishment, use and operation of mitigation banks for the purpose of providing compensation for adverse impacts to wetlands and other aquatic resources. The purpose of this guidance is to clarify the manner in which mitigation banks may be used to satisfy mitigation requirements of the Clean Water Act (CWA) Section 404 permit program and the wetland conservation provisions of the Food Security Act (FSA) (i.e., "Swampbuster" provisions). Recognizing the potential benefits mitigation banking offers for streamlining the permit evaluation process and providing more effective mitigation for authorized impacts to wetlands, the agencies encourage the establishment and appropriate use of mitigation banks in the Section 404 and "Swampbuster" programs.

DATES: The effective date of this Memorandum to the Field is December 28, 1995.

FOR FURTHER INFORMATION CONTACT: Mr. Jack Chowning (Corps) at (202) 761-1781; Mr. Thomas Kelsch (EPA) at (202) 260-8795; Ms.

Sandra Byrd (NRCS) at (202) 690-3501; Mr. Mark Miller (FWS) at (703) 358-2183; Ms. Susan-Marie Stedman (NMFS) at (301) 713-2325.

SUPPLEMENTARY INFORMATION: Mitigating the environmental impacts of necessary development actions on the Nation's wetlands and other aquatic resources is a central premise of Federal wetlands programs. The CWA Section 404 permit program relies on the use of compensatory mitigation to offset unavoidable damage to wetlands and other aquatic resources through, for example, the restoration or creation of wetlands. Under the "Swampbuster" provisions of the FSA, farmers are required to provide mitigation to offset certain conversions of wetlands for agricultural purposes in order to maintain their program eligibility.

Mitigation banking has been defined as wetland restoration, creation, enhancement, and in exceptional circumstances, preservation undertaken expressly for the purpose of compensating for unavoidable wetland losses in advance of development actions, when such compensation cannot be achieved at the development site or would not be as environmentally beneficial. It typically involves the consolidation of small, fragmented wetland mitigation projects into one large contiguous site. Units of restored, created, enhanced or preserved wetlands are expressed as "credits" which may subsequently be withdrawn to offset "debits" incurred at a project development site.

Ideally, mitigation banks are constructed and functioning in advance of development impacts, and are seen as a way of reducing uncertainty in the CWA Section 404 permit program or the FSA "Swampbuster" program by having established compensatory mitigation credit available to an applicant. By consolidating compensation requirements, banks can more effectively replace lost wetland functions within a watershed, as well as provide economies of scale relating to the planning, implementation, monitoring and management of mitigation projects.

On August 23, 1993, the Clinton Administration released a comprehensive package of improvements to Federal wetlands programs which included support for the use of mitigation banks. At that same time, EPA and the Department of the Army issued interim guidance clarifying the role of mitigation banks in the Section 404 permit program and providing general guidelines for their establishment and use. In that document it was acknowledged that additional guidance would be developed, as necessary, following completion of the first phase of the Corps Institute for Water Resources national study on mitigation banking.

The Corps, EPA, NRCS, FWS and NMFS provided notice [60 FR 12286; March 6, 1995] of a proposed guidance on the policy of the Federal government regarding the establishment, use and operation of mitigation banks. The proposed guidance was based, in part, on the experiences to date with mitigation banking, as well as other environmental, economic and institutional issues identified through the Corps national study. Over 130 comments were received on the proposed guidance. The final guidance is based on full and thorough consideration of the public comments received.

A majority of the letters received supported the proposed guidance in general, but suggested modifications to one or more parts of the proposal. In response to these comments, several changes have been made to further clarify the provisions and make other modifications, as necessary, to ensure effective establishment and use of mitigation banks. One key issue on which the agencies received numerous comments focused on the timing of credit withdrawal. In order to provide additional clarification of the changes made to the final guidance in response to comments, the agencies wish to emphasize that it is our intent to ensure that decisions to allow credits to be withdrawn from a mitigation bank in advance of bank maturity be made on a case-by-case basis to best reflect the particular ecological and economic circumstances of each bank. The percentage of advance credits permitted for a particular bank may be higher or lower than the 15 percent example included in the proposed guidance. The final guidance is being revised to eliminate the reference to a specific percentage in order to provide needed flexibility. Copies of the comments and the agencies' response to significant comments are available for public review. Interested parties should contact the agency representatives for additional information.

This guidance does not change the substantive requirements of the Section 404 permit program or the FSA "Swampbuster" program. Rather, it interprets and provides internal guidance and procedures to the agency field personnel for the establishment, use and operation of mitigation banks consistent with existing regulations and policies of each program. The policies set out in this document are not final agency action, but are intended solely as guidance. The guidance is not intended, nor can it be relied upon, to create any rights enforceable by any party in litigation with the United States. This guidance does not establish or affect legal rights or obligations, establish a binding norm on any party and it is not finally determinative of the issues addressed. Any regulatory decisions made by the agencies in any particular matter addressed by this guidance will be made by applying the governing law and regulations to the relevant facts. The purpose of the document is to provide policy and technical guidance to encourage the effective use of mitigation banks as a means of compensating for the authorized loss of wetlands and other aquatic resources.

John H. Zirschky
Acting Assistant Secretary (Civil Works), Department of the Army

Robert Perciasepe
Assistant Administrator for Water, Environmental Protection Agency

Thomas R. Hebert
Acting Undersecretary for Natural Resources and Environment, Department of Agriculture

Robert P. Davison
Acting Assistant Secretary for Fish and Wildlife and Parks, Department of the Interior

Douglas K. Hall
Assistant Secretary for Oceans and Atmosphere, Department of Commerce

Memorandum to the Field

Subject: Federal Guidance for the Establishment, Use and Operation of Mitigation Banks

I. Introduction

A. Purpose and Scope of Guidance

This document provides policy guidance for the establishment, use and operation of mitigation banks for the purpose of providing compensatory mitigation for authorized adverse impacts to wetlands and other aquatic resources. This guidance is provided expressly to assist Federal personnel, bank sponsors, and others in meeting the requirements of Section 404 of the Clean Water Act (CWA), Section 10 of the Rivers and Harbors Act, the wetland conservation provisions of the Food Security Act (FSA) (i.e., "Swampbuster"), and other applicable Federal statutes and regulations. The policies and procedures discussed herein are consistent with current requirements of the Section 10/404 regulatory program and "Swampbuster" provisions and are intended only to clarify the applicability of existing requirements to mitigation banking.

The policies and procedures discussed herein are applicable to the establishment, use and operation of public mitigation banks, as well as privately sponsored mitigation banks, including third party banks (e.g., entrepreneurial banks).

For purposes of this guidance, mitigation banking means the restoration, creation, enhancement and, in exceptional circumstances, preservation of wetlands and/or other aquatic resources expressly for the purpose of providing compensatory mitigation in advance of authorized impacts to similar resources.

The objective of a mitigation bank is to provide for the replacement of the chemical, physical and biological functions of wetlands and other aquatic resources which are lost as a result of authorized impacts. Using appropriate methods, the newly established functions are quantified as mitigation "credits" which are available for use by the bank sponsor or by other parties to compensate for adverse impacts (i.e., "debits"). Consistent with mitigation policies established under the Council on Environmental Quality Implementing Regulations (CEQ regulations) [40 CFR Part 1508.20], and the Section 404(b)(1) Guidelines (Guidelines) [40 CFR Part 230], the use of credits may only be authorized for purposes of complying with Section 10/404 when adverse impacts are unavoidable. In addition, for both the Section 10/404 and "Swampbuster" programs, credits may only be authorized when on-site compensation is either not practicable or use of a mitigation bank is environmentally preferable to on-site compensation. Prospective bank sponsors should not construe or anticipate participation in the establishment of a mitigation bank as ultimate authorization for specific projects, as excepting such projects from any applicable requirements, or as preauthorizing the use of credits from that bank for any particular project.

Mitigation banks provide greater flexibility to applicants needing to comply with mitigation requirements and can have several advantages over individual mitigation projects, some of which are listed below:

1. It may be more advantageous for maintaining the integrity of the aquatic ecosystem to consolidate compensatory mitigation into a single large parcel or contiguous parcels when ecologically appropriate;

2. Establishment of a mitigation bank can bring together financial resources, planning and scientific expertise not practicable to many project-specific compensatory mitigation proposals. This consolidation of resources can increase the potential for the establishment and long-term management of successful mitigation that maximizes opportunities for contributing to biodiversity and/or watershed function;

3. Use of mitigation banks may reduce permit processing times and provide more cost-effective compensatory mitigation opportunities for projects that qualify;

4. Compensatory mitigation is typically implemented and functioning in advance of project impacts, thereby reducing temporal losses of aquatic functions and uncertainty over whether the mitigation will be successful in offsetting project impacts;

5. Consolidation of compensatory mitigation within a mitigation bank increases the efficiency of limited agency resources in the review and compliance monitoring of mitigation projects, and thus improves the reliability of efforts to restore, create or enhance wetlands for mitigation purposes;

6. The existence of mitigation banks can contribute towards attainment of the goal for no overall net loss of the Nation's wetlands by providing opportunities to compensate for authorized impacts when mitigation might not otherwise be appropriate or practicable.

II. Policy Considerations

The following policy considerations provide general guidance for the establishment, use and operation of mitigation banks. It is the agencies' intent that this guidance be applied to mitigation bank proposals submitted for approval on or after the effective date of this guidance and to those in early stages of planning or development. It is not intended that this policy be retroactive for mitigation banks that have already received agency approval. While it is recognized that individual mitigation banking proposals may vary, it is the intent of this guidance that the fundamental precepts be applicable to future mitigation banks.

For the purposes of Section 10/404, and consistent with the CEQ regulations, the Guidelines, and the Memorandum of Agreement Between the Environmental Protection Agency (EPA) and the Department of the Army Concerning the Determination of Mitigation under the Clean Water Act Section 404(b)(1) Guidelines, mitigation means sequentially avoiding impacts, minimizing impacts, and compensating for remaining unavoidable impacts. Compensatory mitigation, under Section 10/404, is the restoration, creation, enhancement, or in exceptional circumstances, preservation of wetlands and/or other aquatic resources for the purpose of compensating for unavoidable adverse impacts. A site where wetlands and/or other aquatic resources are restored, created, enhanced, or in exceptional circumstances, preserved expressly for the purpose of providing compensatory mitigation in advance of authorized impacts to similar resources is a mitigation bank.

A. Authorities

This guidance is established in accordance with the following statutes, regulations, and policies. It is intended to clarify provisions within these existing authorities and does not establish any new requirements.

1. Clean Water Act Section 404 [33 USC 1344].
2. Rivers and Harbors Act of 1899 Section 10 [33 USC 403 et seq.].
3. Environmental Protection Agency, Section 404(b)(1) Guidelines [40 CFR Part 230]. Guidelines for Specification of Disposal Sites for Dredged or Fill Material.

4 Department of the Army, Section 404 Permit Regulations [33 CFR Parts 320-330]. Policies for evaluating permit applications to discharge dredged or fill material.

5. Memorandum of Agreement between the Environmental Protection Agency and the Department of the Army Concerning the Determination of Mitigation under the Clean Water Act Section 404(b)(1) Guidelines [February 6, 1990].
6. Title XII Food Security Act of 1985 as amended by the Food, Agriculture, Conservation and Trade Act of 1990 [16 USC 3801 et seq.].
7. National Environmental Policy Act [42 USC 4321 et seq.], including the Council on Environmental Quality's implementing regulations [40 CFR Parts 1500–1508].
8. Fish and Wildlife Coordination Act [16 USC 661 et seq.].
9. Fish and Wildlife Service Mitigation Policy [46 FR pages 7644–7663, 1981].
10. Magnuson Fishery Conservation and Management Act [16 USC 1801 et seq.].
11. National Marine Fisheries Service Habitat Conservation Policy [48 FR pages 53142–53147, 1983].

The policies set out in this document are not final agency action, but are intended solely as guidance. The guidance is not intended, nor can it be relied upon, to create any rights enforceable by any party in litigation with the United States. This guidance does not establish or affect legal rights or obligations, establish a binding norm on any party and it is not finally determinative of the issues addressed. Any regulatory decisions made by the agencies in any particular matter addressed by this guidance will be made by applying the governing law and regulations to the relevant facts.

B. Planning Considerations

1. Goal setting

The overall goal of a mitigation bank is to provide economically efficient and flexible mitigation opportunities, while fully compensating for wetland and other aquatic resource losses in a manner that contributes to the long-term ecological functioning of the watershed within which the bank is to be located. The goal will include the need to replace essential aquatic functions which are anticipated to be lost through authorized activities within the bank's service area. In some cases, banks may also be used to address other resource objectives that have been identified in a watershed management plan or other resource assessment. It is desirable to set the particular objectives for a mitigation bank (i.e., the type and character of wetlands and/or aquatic resources to be established) in advance of site selection. The goal and objectives should be driven by the anticipated mitigation need; the site selected should support achieving the goal and objectives.

2. Site selection

The agencies will give careful consideration to the ecological suitability of a site for achieving the goal and objectives of a bank, i.e., that it possess the physical, chemical and biological characteristics to support establishment of the desired aquatic resources and functions. Size and location of the site relative to other ecological features, hydrologic sources (including the availability of water rights), and compatibility with adjacent land uses and watershed management plans are important factors for consideration. It also is important that ecologically significant aquatic or upland resources (e.g., shallow sub-tidal habitat, mature forests), cultural sites, or habitat for Federal or state-listed threatened and endangered species are not compromised in the process of establishing a bank. Other significant factors for consideration include, but are not limited to, development trends (i.e., anticipated land use changes), habitat status and trends, local or regional goals for the restoration or protection of particular habitat types or functions (e.g., re-establishment of habitat corridors or habitat for species of concern), water quality and floodplain management goals, and the relative potential for chemical contamination of the wetlands and/or other aquatic resources.

Banks may be sited on public or private lands. Cooperative arrangements between public and private entities to use public lands for mitigation banks may be acceptable. In some circumstances, it may be appropriate to site banks on Federal, state, tribal or locally owned resource management areas (e.g., wildlife management areas, national or state forests, public parks, recreation areas). The siting of banks on such lands may be acceptable if the internal policies of the public agency allow use of its land for such purposes, and the public agency grants approval. Mitigation credits generated by banks of this nature should be based solely on those values in the bank that are supplemental to the public program(s) already planned or in place, that is, baseline values represented by existing or already planned public programs, including preservation value, should not be counted toward bank credits.

Similarly, Federally funded wetland conservation projects undertaken via separate authority and for other purposes, such as the Wetlands Reserve Program, Farmer's Home Administration fee title transfers or conservation easements, and Partners for Wildlife Program, cannot be used for the purpose of generating credits within a mitigation bank. However, mitigation credit may be given for activities undertaken in conjunction with, but supplemental to, such programs in order to maximize the overall ecological benefit of the conservation project.

3. Technical feasibility

Mitigation banks should be planned and designed to be self-sustaining over time to the extent possible. The techniques for establishing wetlands and/or other aquatic resources must be carefully selected, since this science is constantly evolving. The restoration of historic or substantially degraded wetlands and/or other aquatic resources (e.g., prior-converted cropland, farmed wetlands) utilizing proven techniques increases the likelihood of success and typically does not result in the loss of other valuable resources. Thus, restoration should be the first option considered when siting a bank. Because of the difficulty in establishing the correct hydrologic conditions associated with many creation projects and the tradeoff in wetland functions involved with certain enhancement activities, these methods should only be considered where there are adequate assurances to ensure success and that the project will result in an overall environmental benefit.

In general, banks which involve complex hydraulic engineering features and/or questionable water sources (e.g., pumped) are more costly to develop, operate and maintain, and have a higher risk of failure than banks designed to function with little or no human intervention. The former situations should only be considered where there are adequate assurances to ensure success. This guidance recognizes that in some circumstances wetlands must be actively managed to ensure their viability and sustainability. Furthermore, long-term maintenance requirements may be necessary and appropriate in some cases (e.g., to maintain fire-dependent plant communities in the absence of natural fire; to control invasive exotic plant species).

Proposed mitigation techniques should be well understood and reliable. When uncertainties surrounding the technical feasibility of a proposed mitigation technique exist, appropriate arrangements (e.g., financial assurances, contingency plans, additional monitoring requirements) should be in place to increase the likelihood of success. Such arrangements may be phased out or reduced once the attainment of prescribed performance standards is demonstrated.

4. Role of preservation

Credit may be given when existing wetlands and/or other aquatic resources are preserved in conjunction with restoration, creation or enhancement activities, and when it is demonstrated that the preservation will augment the functions of the restored, created or enhanced aquatic resource. Such augmentation may be reflected in the total number of credits available from the bank.

In addition, the preservation of existing wetlands and/or other aquatic resources in perpetuity may be authorized as the sole basis for generating credits in mitigation banks only in exceptional circumstances, consistent with existing regulations, policies and guidance. Under such circumstances, preservation may be accomplished through the implementation of appropriate legal mechanisms (e.g., transfer of deed, deed restrictions, conservation easement) to protect wetlands and/or other aquatic resources, accompanied by implementation of appropriate changes in land use or other physical changes as necessary (e.g., installation of restrictive fencing).

Determining whether preservation is appropriate as the sole basis for generating credits at a mitigation bank requires careful judgment regarding a number of factors. Consideration must be given to whether wetlands and/or other aquatic resources proposed for preservation (1) perform physical or biological functions, the preservation of which is important to the region in which the aquatic resources are located, and (2) are under demonstrable threat of loss or substantial degradation due to human activities that might not otherwise be expected to be restricted. The existence of a demonstrable threat will be based on clear evidence of destructive land use changes which are consistent with local and regional land use trends and are not the consequence of actions under the control of the bank sponsor. Wetlands and other aquatic resources restored under the Conservation Reserve Program or similar programs requiring only temporary conservation easements may be eligible for banking credit upon termination of the original easement if the wetlands are provided permanent protection and it would otherwise be expected that the resources would be converted upon termination of the easement. The number of mitigation credits available from a bank that is based solely on preservation should be based on the functions that would otherwise be lost or degraded if the aquatic resources were not preserved, and the timing of such loss or degradation. As such, compensation for aquatic resource impacts will typically require a greater number of acres from a preservation bank than from a bank which is based on restoration, creation or enhancement.

5. Inclusion of upland areas

Credit may be given for the inclusion of upland areas occurring within a bank only to the degree that such features increase the overall ecological functioning of the bank. If such features are included as part of a bank, it is important that they receive the same protected status as the rest of the bank and be subject to the same operational procedures and requirements. The presence of upland areas may increase the per-unit value of the aquatic habitat in the bank. Alternatively, limited credit may be given to upland areas protected within the bank to reflect the functions inherently provided by such areas (e.g., nutrient and sediment filtration of stormwater runoff, wildlife habitat diversity) which directly enhance or maintain the integrity of the aquatic ecosystem and that might otherwise be subject to threat of loss or degradation. An appropriate functional assessment methodology should be used to determine the manner and extent to which such features augment the functions of restored, created or enhanced wetlands and/or other aquatic resources.

6. Mitigation banking and watershed planning

Mitigation banks should be planned and developed to address the specific resource needs of a particular watershed. Furthermore, decisions regarding the location, type of wetlands and/or other aquatic resources to be established, and proposed uses of a mitigation bank are most appropriately made within the context of a comprehensive watershed plan. Such watershed planning efforts often identify categories of activities having minimal adverse effects on the aquatic ecosystem and that, therefore, could be authorized under a general permit. In order to reduce the potential cumulative effects of such activities, it may be appropriate to offset these types of impacts through the use of a mitigation bank established in conjunction with a watershed plan.

C. Establishment of Mitigation Banks

1. Prospectus

Prospective bank sponsors should first submit a prospectus to the Army Corps of Engineers (Corps) or Natural Resources Conservation Service (NRCS)[1] to initiate the planning and review process by the appropriate agencies. Prior to submitting a prospectus, bank sponsors are encouraged to discuss their proposal with the appropriate agencies (e.g., pre-application coordination).

It is the intent of the agencies to provide practical comments to the bank sponsors regarding the general need for and technical feasibility of proposed banks. Therefore, bank sponsors are encouraged to include in the prospectus sufficient information concerning the objectives for the bank and how it will be established and operated to allow the agencies to provide such feedback. Formal agency involvement and review is initiated with submittal of a prospectus.

2. Mitigation banking instruments

Information provided in the prospectus will serve as the basis for establishing the mitigation banking instrument. All mitigation banks need to have a banking instrument as documentation of agency concurrence on the objectives and administration of the bank. The banking instrument should describe in detail the physical and legal characteristics of the bank, and how the bank will be established and operated. For regional banking programs sponsored by a single entity (e.g., a state transportation agency), it may be appropriate to establish an "umbrella" instrument for the establishment and operation of multiple bank sites. In such circumstances, the need for supplemental site-specific information (e.g., individual site plans) should be addressed in the banking instrument. The banking instrument will be signed by the bank sponsor and the concurring regulatory and resource agencies represented on the Mitigation Bank Review Team (section II.C.2.). The following information should be addressed, as appropriate, within the banking instrument:

a. bank goals and objectives;
b. ownership of bank lands;
c. bank size and classes of wetlands and/or other aquatic resources proposed for inclusion in the bank, including a site plan and specifications;

---

[1]The Corps will typically serve as the lead agency for the establishment of mitigation banks. Bank sponsors proposing establishment of mitigation banks solely for the purpose of complying with the "Swampbuster" provisions of FSA should submit their prospectus to the NRCS.

d. description of baseline conditions at the bank site;
e. geographic service area;
f. wetland classes or other aquatic resource impacts suitable for compensation;
g. methods for determining credits and debits;
h. accounting procedures;
i. performance standards for determining credit availability and bank success;
j. reporting protocols and monitoring plan;
k. contingency and remedial actions and responsibilities;
l. financial assurances;
m. compensation ratios;
n. provisions for long-term management and maintenance.

The terms and conditions of the banking instrument may be amended, in accordance with the procedures used to establish the instrument and subject to agreement by the signatories.

In cases where initial establishment of the mitigation bank involves a discharge into waters of the United States requiring Section 10/404 authorization, the banking instrument will be made part of a Department of the Army permit for that discharge. Submittal of an individual permit application should be accompanied by a sufficiently detailed prospectus to allow for concurrent processing of each. Preparation of a banking instrument, however, should not alter the normal permit evaluation process timeframes. A bank sponsor may proceed with activities for the construction of a bank subsequent to receiving the Department of the Army authorization. It should be noted, however, that a bank sponsor who proceeds in the absence of a banking instrument does so at his/her own risk.

In cases where the mitigation bank is established pursuant to the FSA, the banking instrument will be included in the plan developed or approved by NRCS and the Fish and Wildlife Service (FWS).

3. Agency roles and coordination

Collectively, the signatory agencies to the banking instrument will comprise the Mitigation Bank Review Team (MBRT). Representatives from the Corps, EPA, FWS, National Marine Fisheries Service (NMFS), and NRCS, as appropriate given the projected use for the bank, should typically comprise the MBRT. In addition, it is appropriate for representatives from state, tribal and local regulatory and resource agencies to participate where an agency has authorities and/or mandates directly affecting or affected by the establishment, use or operation of a bank. No agency is required to sign a banking instrument; however, in signing a banking instrument, an agency agrees to the terms of that instrument.

The Corps will serve as Chair of the MBRT, except in cases where the bank is proposed solely for the purpose of complying with the FSA, in which case NRCS will be the MBRT Chair. In addition, where a bank is proposed to satisfy the requirements of another Federal, state, tribal or local program, it may be appropriate for the administering agency to serve as co-Chair of the MBRT.

The primary role of the MBRT is to facilitate the establishment of mitigation banks through the development of mitigation banking instruments. Because of the different authorities and responsibilities of each agency represented on the MBRT, there is a benefit in achieving agreement on the banking instrument. For this reason, the MBRT will strive to obtain consensus on its actions. The Chair of the MBRT will have the responsibility for making final decisions regarding the terms and conditions of the banking instrument where consensus cannot otherwise be reached within a reasonable timeframe (e.g., 90 days from the date of submittal of a complete prospectus). The MBRT will review and seek consensus on the banking instrument and final plans for the restoration, creation, enhancement, and/or preservation of wetlands and other aquatic resources.

Consistent with its authorities under Section 10/404, the Corps is responsible for authorizing use of a particular mitigation bank on a project-specific basis and determining the number and availability of credits required to compensate for proposed impacts in accordance with the terms of the banking instrument. Decisions rendered by the Corps must fully consider review agency comments submitted as part of the permit evaluation process. Similarly, the NRCS, in consultation with the FWS, will make the final decision pertaining to the withdrawal of credits from banks as appropriate mitigation pursuant to FSA.

4. Role of the bank sponsor

The bank sponsor is responsible for the preparation of the banking instrument in consultation with the MBRT. The bank sponsor should, therefore, have sufficient opportunity to discuss the content of the banking instrument with the MBRT. The bank sponsor is also responsible for the overall operation and management of the bank in accordance with the terms of the banking instrument, including the preparation and distribution of monitoring reports and accounting statements/ledger, as necessary.

5. Public review and comment

The public should be notified of and have an opportunity to comment on all bank proposals. For banks which require authorization under an individual Section 10/404 permit or a state, tribal or local program that involves a similar public notice and comment process, this condition will typically be satisfied through such standard procedures. For other proposals, the Corps or NRCS, upon receipt of a complete banking prospectus, should provide notification of the availability of the prospectus for a minimum 21-day public comment period. Notification procedures will be similar to those used by the Corps in the standard permit review process. Copies of all public comments received will be distributed to the other members of the MBRT and the bank sponsor for full consideration in the development of the final banking instrument.

6. Dispute resolution procedure

The MBRT will work to reach consensus on its actions in accordance with this guidance. It is anticipated that all issues will be resolved by the MBRT in this manner.

a. Development of the banking instrument

During the development of the banking instrument, if an agency representative considers that a particular decision raises concern regarding the application of existing policy or procedures, an agency may request, through written notification, that the issue be reviewed by the Corps District Engineer, or NRCS State Conservationist, as appropriate. Said notification will describe the issue in sufficient detail and provide recommendations for resolution. Within 20 days, the District Engineer or State Conservationist (as appropriate) will consult with the notifying agency(ies) and will resolve the issue. The resolution will be forwarded to the other MBRT member agencies. The bank sponsor may also request the District Engineer or State Conservationist review actions taken to develop the banking instrument if the sponsor believes that inadequate progress has been made on the instrument by the MBRT.

b. Application of the banking instrument

As previously stated, the Corps and NRCS are responsible for making final decisions on a project-specific basis regarding the use of a mitigation bank for purposes of Section 10/404 and FSA, respectively. In the event an agency on the MBRT is concerned that a proposed use may be inconsistent with the terms of the banking instrument, that agency may raise the issue to the attention of the Corps or NRCS through the permit evaluation process. In order to facilitate timely and effective consideration of agency comments, the Corps or NRCS, as appropriate, will advise the MBRT agencies of a proposed use of a bank. The Corps will fully consider comments provided by the review agencies regarding mitigation as part of the permit evaluation process. The NRCS will consult with FWS in making its decisions pertaining to mitigation.

If, in the view of an agency on the MBRT, an issued permit or series of permits reflects a pattern of concern regarding the application of the terms of the banking instrument, that agency may initiate review of the concern by the full MBRT through written notification to the MBRT Chair. The MBRT Chair will convene a meeting of the MBRT, or initiate another appropriate forum for communication, typically within 20 days of receipt of notification, to resolve concerns. Any such effort to address concerns regarding the application of a banking instrument will not delay any decision pending before the authorizing agency (e.g., Corps or NRCS).

D. Criteria for Use of a Mitigation Bank

1. Project applicability

All activities regulated under Section 10/404 may be eligible to use a mitigation bank as compensation for unavoidable impacts to wetlands and/or other aquatic resources. Mitigation banks established for FSA purposes may be debited only in accordance with the mitigation and replacement provisions of 7 CFR Part 12.

Credits from mitigation banks may also be used to compensate for environmental impacts authorized under other programs (e.g., state or local wetland regulatory programs, NPDES program, Corps civil works projects, Superfund removal and remedial actions). In no case may the same credits be used to compensate for more than one activity; however, the same credits may be used to compensate for an activity which requires authorization under more than one program.

2. Relationship to mitigation requirements

Under the existing requirements of Section 10/404, all appropriate and practicable steps must be undertaken by the applicant to first avoid and then minimize adverse impacts to aquatic resources, prior to authorization to use a particular mitigation bank. Remaining unavoidable impacts must be compensated to the extent appropriate and practicable. For both the Section 10/404 and "Swampbuster" programs, requirements for compensatory mitigation may be satisfied through the use of mitigation banks when either on-site compensation is not practicable or use of the mitigation bank is environmentally preferable to on-site compensation.

It is important to emphasize that applicants should not expect that establishment of, or purchasing credits from, a mitigation bank will necessarily lead to a determination of compliance with applicable mitigation requirements (i.e., Section 404(b)(1) Guidelines or FSA Manual), or as excepting projects from any applicable requirements.

3. Geographic limits of applicability

The service area of a mitigation bank is the area (e.g., watershed, county) wherein a bank can reasonably be expected to provide appropriate compensation for impacts to wetlands and/or other aquatic resources. This area should be designated in the banking instrument. Designation of the service area should be based on consideration of hydrologic and biotic criteria, and be stipulated in the banking instrument. Use of a mitigation bank to compensate for impacts beyond the designated service area may be authorized, on a case-by-case basis, where it is determined to be practicable and environmentally desirable.

The geographic extent of a service area should, to the extent environmentally desirable, be guided by the cataloging unit of the "Hydrologic Unit Map of the United States" (USGS, 1980) and the ecoregion of the "Ecoregions of the United States" (James M. Omernik, EPA, 1986) or section of the "Descriptions of the Ecoregions of the United States" (Robert G. Bailey, USDA, 1980). It may be appropriate to use other classification systems developed at the state or regional level for the purpose of specifying bank service areas, when such systems compare favorably in their objectives and level of detail. In the interest of integrating banks with other resource management objectives, bank service areas may encompass larger watershed areas if the designation of such areas is supported by local or regional management plans (e.g. Special Area Management Plans, Advance Identification), State Wetland Conservation Plans or other Federally sponsored or recognized resource management plans. Furthermore, designation of a more inclusive service area may be appropriate for mitigation banks whose primary purpose is to compensate for linear projects that typically involve numerous small impacts in several different watersheds.

4. Use of a mitigation bank vs. on-site mitigation

The agencies' preference for on-site mitigation, indicated in the 1990 Memorandum of Agreement on mitigation between the EPA and the Department of the Army, should not preclude the use of a mitigation bank when there is no practicable opportunity for on-site compensation, or when use of a bank is environmentally preferable to on-site compensation. On-site mitigation may be preferable where there is a practicable opportunity to compensate for important local functions including local flood control functions, habitat for a species or population with a very limited geographic range or narrow environmental requirements, or where local water quality concerns dominate.

In choosing between on-site mitigation and use of a mitigation bank, careful consideration should be given to the likelihood for successfully establishing the desired habitat type, the compatibility of the mitigation project with adjacent land uses, and the practicability of long-term monitoring and maintenance to determine whether the effort will be ecologically sustainable, as well as the relative cost of mitigation alternatives. In general, use of a mitigation bank to compensate for minor aquatic resource impacts (e.g., numerous, small impacts associated with linear projects, impacts authorized under nationwide permits) is preferable to on-site mitigation. With respect to larger aquatic resource impacts, use of a bank may be appropriate if it is capable of replacing essential physical and/or biological functions of the aquatic resources which are expected to be lost or degraded. Finally, there may be circumstances warranting a combination of on-site and off-site mitigation to compensate for losses.

5. In-kind vs. out-of-kind mitigation determinations

In the interest of achieving functional replacement, in-kind compensation of aquatic resource impacts should generally be required. Out-of-kind compensation may be acceptable if it is determined to be practicable and environmentally preferable to in-kind compensation (e.g., of greater ecological value to a particular region). However, nontidal wetlands should typically not be used to compensate for the loss or degradation of tidal wetlands. Decisions regarding out-of-kind mitigation are typically made on a case-by-case basis during the permit evaluation process. The banking instrument may identify circumstances in which it is environmentally desirable to allow out-of-kind compensation within the context of a particular mitigation bank (e.g., for banks restoring a complex of associated wetland types). Mitigation banks developed as part of an area-wide management plan to address a specific resource objective (e.g. restoration of a particularly vulnerable or valuable wetland habitat type) may be such an example.

6. Timing of credit withdrawal

The number of credits available for withdrawal (i.e., debiting) should generally be commensurate with the level of aquatic functions attained at a bank at the time of debiting. The level of function may be determined through the application of performance standards tailored to the specific restoration, creation or enhancement activity at the bank site or through the use of an appropriate functional assessment methodology.

The success of a mitigation bank with regard to its capacity to establish a healthy and fully functional aquatic system relates directly to both the ecological and financial stability of the bank. Since financial considerations are particularly critical in early stages of bank development, it is generally appropriate, in cases where there is adequate financial assurance and where the likelihood of success of the bank is high, to allow limited debiting of a percentage of the total credits projected for the bank at maturity. Such determinations should take into consideration the initial capital costs needed to establish the bank, and the likelihood of its success. However, it is the intent of this policy to ensure that those actions necessary for the long-term viability of a mitigation bank be accomplished prior to any debiting of the bank. In this regard, the following minimum requirements should be satisfied prior to debiting: (1) banking instrument and mitigation plans have been approved; (2) bank site has been secured; and (3) appropriate financial assurances have been established. In addition, initial physical and biological improvements should be completed no later than the first full growing season following initial debiting of a bank. The temporal loss of functions associated with the debiting of projected credits may justify the need for requiring higher compensation ratios in such cases. For mitigation banks which propose multiple-phased construction, similar conditions should be established for each phase.

Credits attributed to the preservation of existing aquatic resources may become available for debiting immediately upon implementation of appropriate legal protection accompanied by appropriate changes in land use or other physical changes, as necessary.

7. Crediting/debiting/accounting procedures

Credits and debits are the terms used to designate the units of trade (i.e., currency) in mitigation banking. Credits represent the accrual or attainment of aquatic functions at a bank; debits represent the loss of aquatic functions at an impact or project site. Credits are debited from a bank when they are used to offset aquatic resource impacts (e.g., for the purpose of satisfying Section 10/404 permit or FSA requirements).

An appropriate functional assessment methodology (e.g., Habitat Evaluation Procedures, hydrogeomorphic approach to wetlands functional assessment, other regional assessment methodology) acceptable to all signatories should be used to assess wetland and/or other aquatic resource restoration, creation and enhancement activities within a mitigation bank, and to quantify the amount of available credits. The range of functions to be assessed will depend upon the assessment methodology identified in the banking instrument. The same methodology should be used to assess both credits and debits. If an appropriate functional assessment methodology is impractical to employ, acreage may be used as a surrogate for measuring function. Regardless of the method employed, the number of credits should reflect the difference between site conditions under the with- and without-bank scenarios.

The bank sponsor should be responsible for assessing the development of the bank and submitting appropriate documentation of such assessments to the authorizing agency(ies), who will distribute the documents to the other members of the MBRT for review. Members of the MBRT are encouraged to conduct regular (e.g., annual) on-site inspections, as appropriate, to monitor bank performance. Alternatively, functional assessments may be conducted by a team representing involved resource and regulatory agencies and other appropriate parties. The number of available credits in a mitigation bank may need to be adjusted to reflect actual conditions.

The banking instrument should require that bank sponsors establish and maintain an accounting system (i.e., ledger) which documents the activity of all mitigation bank accounts. Each time an approved debit/credit transaction occurs at a given bank, the bank sponsor should submit a statement to the authorizing agency(ies). The bank sponsor should also generate an annual ledger report for all mitigation bank accounts to be submitted to the MBRT Chair for distribution to each member of the MBRT.

Credits may be sold to third parties. The cost of mitigation credits to a third party is determined by the bank sponsor.

8. Party responsible for bank success

The bank sponsor is responsible for assuring the success of the debited restoration, creation, enhancement and preservation activities at the mitigation bank, and it is therefore extremely important that an enforceable mechanism be adopted establishing the responsibility of the bank sponsor to develop and operate the bank properly. Where authorization under Section 10/404 and/or FSA is necessary to establish the bank, the Department of the Army permit or NRCS plan should be conditioned to ensure that provisions of the banking instrument are enforceable by the appropriate agency(ies). In circumstances where establishment of a bank does not require such authorization, the details of the bank sponsor's responsibilities should be delineated by the relevant authorizing agency (e.g., the Corps in the case of Section 10/404 permits) in any permit in which the permittee's mitigation obligations are met through use of the bank. In addition, the bank sponsor should sign such permits for the limited purpose of meeting those mitigation responsibilities, thus confirming that those responsibilities are enforceable against the bank sponsor if necessary.

E. Long-Term Management, Monitoring and Remediation

1. Bank operational life

The operational life of a bank refers to the period during which the terms and conditions of the banking instrument are in effect. With the exception of arrangements for the long-term management and protection in perpetuity of the wetlands and/or other aquatic resources, the operational life of a mitigation bank terminates at the point when (1) compensatory mitigation credits have been exhausted or banking activity is voluntarily terminated with written notice by the bank sponsor provided to the Corps or NRCS and other members of the MBRT, and (2) it has been determined that the debited bank is functionally mature and/or self-sustaining to the degree specified in the banking instrument.

2. Long-term management and protection

The wetlands and/or other aquatic resources in a mitigation bank should be protected in perpetuity with appropriate real estate arrangements (e.g., conservation easements, transfer of title to Federal or State resource agency or non-profit conservation organization). Such arrangements should effectively restrict harmful activities (i.e., incompatible uses[2]) that might otherwise jeopardize the purpose of the bank. In exceptional circumstances, real estate arrangements may be approved which dictate finite protection for a bank (e.g., for coastal protection projects which prolong the ecological viability of the aquatic system). However, in no case should finite protection extend for a lesser time than the duration of project impacts for which the bank is being used to provide compensation.

The bank sponsor is responsible for securing adequate funds for the operation and maintenance of the bank during its operational life, as well as for the long-term management of the wetlands and/or other aquatic resources, as necessary. The banking instrument should identify the entity responsible for the ownership and long-term management of the wetlands and/or other aquatic resources. Where needed, the acquisition and protection of water rights should be secured by the bank sponsor and documented in the banking instrument.

3. Monitoring requirements

The bank sponsor is responsible for monitoring the mitigation bank in accordance with monitoring provisions identified in the banking instrument to determine the level of success and identify problems requiring remedial action. Monitoring provisions should be set forth in the banking instrument and based on scientifically sound performance standards prescribed for the bank. Monitoring should be conducted at time intervals appropriate for the particular project type and until such time that the authorizing agency(ies), in consultation with the MBRT, are confident that success is being achieved (i.e., performance standards are attained). The period for monitoring will typically be five years; however, it may be necessary to extend this period for projects requiring more time to reach a stable condition (e.g., forested wetlands) or where remedial activities were undertaken. Annual monitoring reports should be submitted to the authorizing agency(ies), who is responsible for distribution to the other members of the MBRT, in accordance with the terms specified in the banking instrument.

4. Remedial action

The banking instrument should stipulate the general procedures for identifying and implementing remedial measures at a bank, or any portion thereof. Reme-

[2]For example, certain silvicultural practices (e.g., clear cutting and/or harvests on short-term rotations) may be incompatible with the objectives of a mitigation bank. In contrast, silvicultural practices such as long-term rotations, selective cutting, maintenance of vegetation diversity, and undisturbed buffers are more likely to be considered a compatible use.

dial measures should be based on information contained in the monitoring reports (i.e., the attainment of prescribed performance standards), as well as agency site inspections. The need for remediation will be determined by the authorizing agency(ies) in consultation with the MBRT and bank sponsor.

5. Financial assurances

The bank sponsor is responsible for securing sufficient funds or other financial assurances to cover contingency actions in the event of bank default or failure. Accordingly, banks posing a greater risk of failure and where credits have been debited, should have comparatively higher financial sureties in place, than those where the likelihood of success is more certain. In addition, the bank sponsor is responsible for securing adequate funding to monitor and maintain the bank throughout its operational life, as well as beyond the operational life if not self-sustaining. Total funding requirements should reflect realistic cost estimates for monitoring, long-term maintenance, contingency and remedial actions.

Financial assurances may be in the form of performance bonds, irrevocable trusts, escrow accounts, casualty insurance, letters of credit, legislatively enacted dedicated funds for government-operated banks or other approved instruments. Such assurances may be phased out or reduced, once it has been demonstrated that the bank is functionally mature and/or self-sustaining (in accordance with performance standards).

F. Other Considerations

1. In-lieu-fee mitigation arrangements

For purposes of this guidance, in-lieu-fee, fee mitigation, or other similar arrangements, wherein funds are paid to a natural resource management entity for implementation of either specific or general wetland or other aquatic resource development projects, are not considered to meet the definition of mitigation banking because they do not typically provide compensatory mitigation in advance of project impacts. Moreover, such arrangements do not typically provide a clear timetable for the initiation of mitigation efforts. The Corps, in consultation with the other agencies, may find there are circumstances where such arrangements are appropriate so long as they meet the requirements that would otherwise apply to an off-site, prospective mitigation effort and provides adequate assurances of success and timely implementation. In such cases, a formal agreement between the sponsor and the agencies, similar to a banking instrument, is necessary to define the conditions under which its use is considered appropriate.

2. Special considerations for "Swampbuster"

Current FSA legislation limits the extent to which mitigation banking can be used for FSA purposes. Therefore, if a mitigation bank is to be used for FSA purposes, it must meet the requirements of FSA.

III. Definitions

For the purposes of this guidance document the following terms are defined:

A. *authorizing agency.* Any Federal, state, tribal or local agency that has authorized a particular use of a mitigation bank as compensation for an authorized activity; the authorizing agency will typically have the enforcement authority to ensure that the terms and conditions of the banking instrument are satisfied.

B. *bank sponsor.* Any public or private entity responsible for establishing and, in most circumstances, operating a mitigation bank.

C. *compensatory mitigation.* For purposes of Section 10/404, compensatory mitigation is the restoration, creation, enhancement, or in exceptional circumstances, preservation of wetlands and/or other aquatic resources for the purpose of compensating for unavoidable adverse impacts which remain after all appropriate and practicable avoidance and minimization has been achieved.

D. *consensus.* The term consensus, as defined herein, is a process by which a group synthesizes its concerns and ideas to form a common collaborative agreement acceptable to all members. While the primary goal of consensus is to reach agreement on an issue by all parties, unanimity may not always be possible.

E. *creation.* The establishment of a wetland or other aquatic resource where one did not formerly exist.

F. *credit.* A unit of measure representing the accrual or attainment of aquatic functions at a mitigation bank; the measure of function is typically indexed to the number of wetland acres restored, created, enhanced or preserved.

G. *debit.* A unit of measure representing the loss of aquatic functions at an impact or project site.

H. *enhancement.* Activities conducted in existing wetlands or other aquatic resources which increase one or more aquatic functions.

I. *mitigation.* For purposes of Section 10/404 and consistent with the Council on Environmental Quality regulations, the Section 404(b)(1) Guidelines and the Memorandum of Agreement Between the Environmental Protection Agency and the Department of the Army Concerning the Determination of Mitigation under the Clean Water Act Section 404(b)(1) Guidelines, mitigation means sequentially avoiding impacts, minimizing impacts, and compensating for remaining unavoidable impacts.

J. *mitigation bank.* A mitigation bank is a site where wetlands and/or other aquatic resources are restored, created, enhanced, or in exceptional circumstances, preserved expressly for the purpose of providing compensatory mitigation in advance of authorized impacts to similar resources. For purposes of Section 10/404, use of a mitigation bank may only be authorized when impacts are unavoidable.

K. *Mitigation Bank Review Team (MBRT).* An interagency group of Federal, state, tribal and/or local regulatory and resource agency representatives which are signatory to a banking instrument and oversee the establishment, use and operation of a mitigation bank.

L. *practicable.* Available and capable of being done after taking into consideration cost, existing technology, and logistics in light of overall project purposes.

M. *preservation.* The protection of ecologically important wetlands or other aquatic resources in perpetuity through the implementation of appropriate legal and physical mechanisms. Preservation may include protection of upland areas adjacent to wetlands as necessary to ensure protection and/or enhancement of the aquatic ecosystem.

N. *restoration.* Re-establishment of wetland and/or other aquatic resource characteristics and function(s) at a site where they have ceased to exist, or exist in a substantially degraded state.

O. *service area.* The service area of a mitigation bank is the designated area (e.g., watershed, county) wherein a bank can reasonably be expected to provide appropriate compensation for impacts to wetlands and/or other aquatic resources.

John H. Zirschky
Acting Assistant Secretary (Civil Works), Department of the Army

Robert Perciasepe
Assistant Administrator for Water, Environmental Protection Agency

Thomas R. Hebert
Acting Undersecretary for Natural Resources and Environment, Department of Agriculture

Robert P. Davison
Acting Assistant Secretary for Fish and Wildlife and Parks, Department of the Interior

Douglas K. Hall
Assistant Secretary for Oceans and Atmosphere, Department of Commerce

M[illegible]ation. [illegible] other aquatic resources [illegible] through [illegible] and physical manipulations. Preservation may include [illegible] of upland areas adjacent to wetlands as necessary to ensure protection and/or enhancement of the aquatic ecosystem.

R[illegible]oration. Re-establishment of [illegible] characteristics and functions at a site where they have ceased to exist, or exist in a substantially degraded state.

[illegible] area. The service area of a mitigation bank is the designated area (e.g., watershed, county) wherein a bank can reasonably be expected to provide appropriate compensation for impacts to wetlands and other aquatic resources.

John M. Z[illegible]
[illegible]

Robert [illegible]
[illegible]

Thomas [illegible]
[illegible]

Robert [illegible]
[illegible]

Douglas R. Hall
[illegible]

Bibliography

Adamus, Paul R., et al. 1987. *Wetland Evaluation Technique (WET), Volume II—Methodology.* Vicksburg, MS: U.S. Army Corps of Engineers, Waterways Experiment Station.

Anderson, Robert, and Robert DeCaprio. 1992. "Banking on the Bayou." *National Wetlands Newsletter* 14(1): 10.

Anderson, Robert, and Mark Rockel. April 1991. *Economic Valuation of Wetlands: Discussion Paper #065.* Washington, DC: American Petroleum Institute.

Association of State Wetlands Managers. "Mitigation Banks and Joint Projects in the Context of Wetland Management Plans." Proceedings from a National Wetland Symposium, June 24–27, 1992, Palm Beach Gardens, Florida.

Austin, Jay, James McElfish, and Sara Nicholas. 1993. *Wetland Mitigation Banking.* Washington, DC: Environmental Law Institute.

Bierly, Ken. 1987. "Oregon Mitigation Banking." In *Proceedings, Northwest Wetlands: What Are They? For Whom? For What?* pp. 197–200. Seattle: Institute for Environmental Studies, University of Washington.

Brown, J. H. 1978. "The theory of insular biogeography and the distribution of boreal birds and mammals." *Great Basin Nat. Mem.* 2: 209–222

Brown, J. H. 1971. "Mammals on mountain tops. Non equilibrium insular biogeography." *American Naturalist* 105: 467–478.

Brown, M., and J. J. Dinsmore. 1986. "Implications of marsh size and isolation for marsh bird management." *Journal of Wildlife Management* 50: 392–397.

CH2M HILL, King and Associates. 1994. *Assessing Environmental Benefits and Creating Wetland Mitigation Opportunities for Water Supply Projects.* American Water Works Association, Water Industry Technical Action Fund.

Chesapeake Bay Executive Council. 1988. "Population Growth and Development in the Chesapeake Bay Watershed to the Year 2020." Report of the Year 2020 Panel of Experts to the Chesapeake Executive Council.

City of Eugene, Public Works Engineering. 1995. *Eastern Gateway Wetlands Restoration Project: 1994 Annual Report.* Eugene, Oregon.

Clark, Darryl R. 1990. "Mitigation Banking in Coastal Louisiana: General Banking Procedures and MOA Provisions." Paper presented by Louisiana Department of Natural Resources, Baton Rouge, at the State Wetland Managers Conference, Jackson, Mississippi, April 1990.

Diamond, J. M. 1975. "The island dilemma: Lessons of modern biogeographic studies for the design of nature preserves." *Biological Conservation* 7: 129–145.

DuPriest, Douglas M., and Jon Christenson. 1988. "Constraints on Mitigation Banking: Oregon's Mitigation Banking Act of 1987." *National Wetlands Newsletter* 10(6): 9–11.

Federal Guidance for the Establishment, Use and Operation of Mitigation Banks, *Federal Register* Vol 60, No. 43, Monday, March 6, 1995.

Galli, A. E., C. F. Leck, and R. T. Forman. 1976. "Avian distribution patterns in forested islands of different sizes in central New Jersey." *Auk* 93: 356–365.

Gibbs, J. P., and S. M. Melvin. 1989. "An Assessment of Wading Birds and Other Wetlands Avifauna and Their Habitats in Maine." Final Report Maine Department Inland Fisheries and Wildlife Bangor, Maine. p. 114.

Grenell, Peter, and Melanie Denninger. 1992. *Banks and Joint Projects.* Oakland: California State Coastal Conservancy.

Haynes, William J., II, and Royal C. Gardner. May 1993. "The Value of Wetlands as Wetlands: The Case for Mitigation Banking." *Environmental Law Reporter* 23(50): 10261–10265.

Institute of Water Resources, U.S. Army Corps of Engineers. "National Wetland Mitigation Banking Study: Wetland Mitigation Banking Concepts," prepared by Richard Reppert, IWR Report 94-WMB-1, Alexandria, VA, July 1992.

Institute of Water Resources, U.S. Army Corps of Engineers. "National Wetland Mitigation Banking Study: Expanding Opportunities for Successful Mitigation—The Private Credit Market Alternative," prepared by Leonard Shabman, Paul Scodari, and Dennis King, IWR Report 94-WMB-3, Alexandria, VA, July 1992.

Institute of Water Resources, U.S. Army Corps of Engineers. "National Wetland Mitigation Banking Study: Wetland Mitigation Banking: Resource Document," IWR Report 94-WMB-2, Alexandria, VA, January 1994.

Institute of Water Resources, U.S. Army Corps of Engineers. "National Wetland Mitigation Banking Study: First Phase Report," IWR Report 94-WMB-4, Alexandria, VA, January 1994.

Institute of Water Resources, U.S. Army Corps of Engineers. "National Wetland Mitigation Banking Study: Wetland Mitigation Banking," prepared by Environmental Law Institute, IWR Report 94-WMB-6, Alexandria, VA, February 1994.

Institute of Water Resources, U.S. Army Corps of Engineers. "National Wetland Mitigation Banking Study: An Examination of Wetland Programs: Opportunities for Compensatory Mitigation," prepared by Apogee Research, Inc., IWR Report 94-WMB-5, Alexandria, VA, March 1994.

King, Dennis M. 1992. "The Economics of Ecological Restoration." In J. Duffield and K. Ward, (eds.) *Natural Resource Damages: Law and Economics.* New York: John Wiley & Sons.

King, Dennis M. 1992. "Avoiding Another Taxpayer Bailout." *National Wetlands Newsletter* 14(1): 11–12.

King, Dennis M., and Curtis C. Bohlen. 1994. *Making Sense of Wetland Restoration Costs.* University of MD: Center for Environmental and Estuarine Studies.

Kusler, Jon. 1992. "The Mitigation Banking Debate." *National Wetlands Newsletter* 14(1):4.

MacArthur, R. H., and E. O. Wilson. 1967. *The Theory of Island Biogeography.* Princeton, New Jersey: Princeton University Press.

Marsh, Lindell L., and Dennis R. Acker. 1992. "Mitigation Banking on a Wider Plane." *National Wetlands Newsletter* 14(1): 8–9.

Matthiae, P. E., and F. Stearns. 1981. "Mammals in forest islands in Southeastern Wisconsin." In R. L. Burgess and D. M. Sharpe, (eds.) *Forest Islands Dynamics in Man-Dominated Landscapes, Ecological Series,* pp. 55–56, vol. 41. New York: Springer-Verlag.

National Research Council. 1992. *Restoration of Aquatic Ecosystems.* Washington, DC, National Academy Press.

Riddle, Elizabeth P. 1988. "Mitigation Banks: Unmitigated Disaster or Sound Investment?" In Jon A. Kusler, Millicent L. Quammen, and Gail Brooks, (eds.) *Proceedings of the National Wetland Symposium: Mitigation of Impacts and Losses,* pp. 353–358, Technical Report Number 3. New Orleans, Louisiana: Association of State Wetland Managers, Inc.

Robbins, C. S. 1979. "Effect of forest fragmentation on bird populations." In R. M. Degraaf and K. E. Evans (compilers) *Proceedings of the Workshop: Management of North Central and Northeastern Forests for Nongame Birds,* pp. 198–213. General Tech. Rep NC-51. U.S.D.A. Forest Service St. Paul, MN.

Rogers, L. L., and A. W. Allan. 1987. "Habitat suitability index models: black bear, Upper Great Lakes region." *Biological Report 82.* Washington, DC: U.S. Fish and Wildlife Service. 50 pp.

Salvesen, David. 1994. *Wetlands: Mitigating and Regulating Development Impacts.* Washington, DC: The Urban Land Institute.

Salvesen, David. 1995. "Banking On Wetlands." *Planning* 61(2): 11–15.

Salvesen, David. 1993. "Banking On Wetlands." *Urban Land* 52(6): 36–40.

Shirey, P. "Regional Plans and Mitigation Banking: An Oregon Example." Presented at Wetlands in Washington Conference, Professional Education Systems, Inc., Eau Claire, WI. 1991.

Short, Cathleen. 1988. *Mitigation Banking*. Biological Report 88(41) 103 pp. Washington, DC: U.S. Department of the Interior, Fish and Wildlife Service, Research and Development.

Soileau, D. 1984. "Final Report on the Tenneco LaTerre Corporation Mitigation Banking Proposal, Terrebonne Parish, Louisiana," prepared for the U.S. Fish and Wildlife Service, Division of Ecological Services, Lafayette, LA.

Sokolove, Robert D., and Pamela D. Huang. 1992. "Privatization of Wetland Mitigation Banking." *Natural Resources and Environment* 7(1):36.

Terborgh, John. 1992. "Why American Songbirds are Vanishing." *Scientific American* 266(5): 98–105.

U.S. Department of Interior. 1995. *Endangered Ecosystems of the United States: A Preliminary Assessment of Ecosystems Loss and Degradation*. National Biological Survey, Biological Report 28. Washington, DC: U.S. Department of Interior.

U.S. Forest Service. 1991. *Riparian Forest Buffers*. NA-PR-07-91. Radnor, PA: Forest Resources Management.

U.S. Environmental Protection Agency. 1987. *Unfinished Business: A Comparative Assessment of Environmental Risk*. Washington, DC: Environmental Protection Agency.

U.S. Environmental Protection Agency and the Department of the Army (August 23, 1993, Memorandum to the Field: Establishment and Use of Wetland Mitigation Banks in the Clean Water Act Section 404 Regulatory Program.

U.S. Fish and Wildlife Service. 1980. "Habitat Evaluation Procedures." *Ecological Services Manual, 102-ESM1,* Washington, DC: Department of the Interior.

Weller, M. W., and L. H. Fredrickson. 1974. "Avian ecology of a managed glacial marsh." *Living Bird* 12: 269–291.

Whitcomb, R. F., C. S. Robbins, J. F. Lunch, B. L.. Whitcomb, M. K. Klimkiewicz, and D. Bystrak. 1981. "Effects of forest fragmentation on avifauna of the eastern deciduous forest." In R. L. Burgess and D. M. Sharpe, (eds.) *Forest Island Dynamics in Man-Dominated Landscapes,* pp. 123–205. New York: Springer-Verlag.

Wilkey, P. L. et al. 1994. *Wetlands Mitigation Banking for the Oil and Gas Industry: Assessment, Conclusions, and Recommendations.* ANL/ESD/TM-

63. Argonne National Laboratory, U.S. Department of Energy, Argonne, Illinois.

Wilson, E. O. 1988. *Biodiversity.* Washington, DC: National Academy Press.

World Wildlife Fund. Draft 10/91. "Managing Private Lands to Conserve Biodiversity," Washington, DC.

World Wildlife Fund. "Mitigation Banking: The Pros and Cons," In World Wildlife Fund, *Statewide Wetlands Strategies: A Guide to Protecting and Managing the Resource,* p.68. Washington, DC: Island Press. 1992.

Contributors

Virginia Albrecht is a partner in Beveridge & Diamond's Washington, D.C., office, practicing with the firm's wetlands and endangered species group. Ms. Albrecht is on the associate faculty of the Lincoln Institute of Land Policy and a trustee of the Eastern Mineral Law Foundation.

Cameron Barrows is the southern California area director for The Nature Conservancy, where he has worked for 15 years, including 10 years as the principal TNC staff responsible for the implementation of the Coachella Valley fringe-toed lizard Habitat Conservation Plan. Mr. Barrows is the author of numerous scientific and popular articles dealing with conservation biology.

Terrie Bates is responsible for all regulatory functions of the South Florida Water Management District, including permitting, compliance, and enforcement. For the last two years, she has worked on the interagency team responsible for developing Florida's wetland mitigation banking and environmental resource use permit rules.

Frank Bernadino is the Assistant to the Director of the Department of Environmental Resources Management, Dade County, Florida. He has over 11 years experience in research, regulatory, and planning in south Florida. Mr. Bernardino is one of the principals responsible for the first mitigation bank approved under Florida's mitigation banking regulations.

Alex Camacho is a legal intern with Siemon, Larsen & Marsh in Irvine, California, where he has worked on endangered species issues. Mr. Camacho has been admitted to the University of California school of law. He has a bachelor of arts degree from the University of California–Irvine.

Jan Goldman-Carter is an environmental lawyer and consultant in West Chester, Pennsylvania. She works primarily in the areas of water quality and wetlands protection for environmental groups and governmental agencies. Ms. Goldman-Carter has taught environmental law as an adjunct professor, has written extensively on wetland issues, and was counsel for fisheries and wildlife at the National Wildlife Federation in Washington, D.C.

Steve Gordon is principal planner with the Lane Council of Governments (LCOG) in Eugene, Oregon. He has worked at LCOG since 1975. Since 1987, he has worked on the west Eugene wetlands program. In 1992, Mr. Gordon received the National Wetlands Protection Award from the Environmental Law Institute and the Environmental Protection Agency for his outstanding achievements in conserving wetlands.

Dennis King is the founder and director of King and Associates, Inc., a Washington-based environmental and economic consulting firm and a research professor at the University of Maryland, Center for Environmental and Estuarine Studies. Dr. King has over 25 years of applied research and consulting experience dealing with the interaction of business, trade, and the environment.

Lindell L. Marsh is a partner with the law firm of Siemon, Larsen & Marsh in Irvine, California, practicing in California and Washington D.C. He specializes in collaborative planning to resolve conservation/development issues. A significant area of his practice focuses on wetlands- and wildlife-related issues.

Grady McCallie is a wetlands lobbyist with the National Wildlife Federation in Washington, D.C. Before joining the National Wildlife Federation, Mr. McCallie worked for the Chesapeake Bay Foundation in Annapolis, Maryland.

James M. McElfish Jr. is a senior attorney with the Environmental Law Institute, where he focuses on enforcement policy, natural resources and property law, mining, wetlands, and hazardous waste issues. He has written extensively on state environmental laws, wetlands, and mining and has worked for a law firm and for the U.S. Department of the Interior.

Sara Nicholas is the Director of the Wetlands and Private Lands Initiative at the National Fish and Wildlife Foundation, a private, nonprofit foundation that awards matching grants for conservation projects. Previously, she edited the *National Wetlands Newsletter* for the Environmental Law Institute and coauthored a study on mitigation banking with James McElfish for the EPA and Corps of Engineers.

Douglas R. Porter is President of the Growth Management Institute and is a planning and development consultant in Chevy Chase, Maryland. Mr. Porter is a nationally recognized expert in growth management and has written and spoken widely on this and other issues. His latest book, *Collaborative Planning for Wetlands and Wildlife*, which he coedited with David Salvesen, was published in 1995 by Island Press.

Ann Redmond has been the mitigation coordinator for the Florida Department of Environmental Protection since 1986. She has previous experience with the Northwest Florida Water Management District and has worked as an environmental consultant.

Robert M. Rhodes is a partner in the Florida law firm of Steel Hector & Davis in Tallahassee, where he heads the firm's Environmental and Land Use Department. He chaired the state of Florida's Environmental Regulation Commission Mitigation Banking Task Force, a source for much of Florida's mitigation banking program.

John W. Rogers is a senior vice president at CH2M Hill and a nationally recognized expert in ecosystem and watershed management and in decision science. Mr. Rogers facilitated the decision-making process for the panel of experts convened under the Chesapeake Bay Agreement and assisted in developing the research for the Freshwater Wetlands buffer legislation for the state of New Jersey.

David A. Salvesen is an environmental writer and consultant in Kensington, Maryland. He has consulted and written widely on issues such as wetlands mitigation, watershed management, and endangered species protection. His latest book, *Collaborative Planning for Wetlands and Wildlife,* which he coedited with Douglas Porter, was published in 1995 by Island Press.

Paul Scodari is an economist and project manager with King & Associates, Inc., an environmental economics consulting firm in Washington, D.C. Mr. Scodari has twelve years of experience directing economic analysis for the U.S. Environmental Protection Agency, the U.S. Army Corps of Engineers, and other agencies, as well as private sector clients. He has published numerous articles, books, and monographs on resource economics.

Leonard A. Shabman is a professor of environmental and resource economics in the Department of Agricultural and Applied Economics, Virginia Tech, Blacksburg, Virginia, and a private consultant. He has served as visiting scholar with the U.S. Water Resources Council and Scientific Advisor, Office of the Assistant Secretary of Army for Civil Works, and as a member of the National Research Council Committee on the Restoration of Aquatic Ecosystems.

Robert D. Sokolove serves as President of U.S. Wetland Services, Inc., which provides wetland permitting, mitigation, and mitigation banking services nationwide. He also directs the environmental law firm of Sokolove & Associates. Both companies are located in Washington, D.C. Mr. Sokolove is a regular speaker at industry and trade association meetings and conventions and has published numerous articles.

W. Brooks Stillwell is an attorney in Savannah, Georgia, and counsel to W.E.T., inc. He is a partner in the firm of Hunter, Maclean, Exley & Dunn, P.C., a past Chair of the Real Property Law Section of the State Bar of Georgia, and a member of the American College of Real Estate Lawyers.

John Studt is head of the U.S. Army Corps of Engineers Regulatory Program in south Florida. Prior to his current position he was chief of the Corps Headquarters Regulatory Branch in Washington, D.C., and also worked for two years with the U.S. Environmental Protection Agency's Office of Wetlands Protection.

John M. Tettemer is President of John M. Tettemer & Associates, Ltd. (JMTA), a Costa Mesa–based environmental engineering firm. Founded in 1978, JMTA specializes in conceptual planning and design for projects involving wetlands, water quality, river engineering, and regulatory approvals. Prior to founding JMTA, Mr. Tettemer worked for 24 years with the Los Angeles County Flood Control District.

Michelle Wenzel is an associate in Beveridge & Diamond's Washington, D.C. office, practicing primarily with the firm's wetlands/endangered species and water practice groups. Prior to joining Beveridge & Diamond, she worked as a software engineer for Martin Marietta Astronautics Group and Litton Guidance & Control Systems.

Jora Young is the Director of Science and Stewardship for the Florida Region of The Nature Conservancy, where she oversees management, restoration, and research activities on a network of 40 private sanctuaries and coordinates the development of partnerships with other resource managers throughout the state. Jora designed and supervised the development of an 11,500-acre mitigation project called the Disney Preserve and has been active in conservation land management for over 20 years.

Index